AF362235

Soil Mechanics Fundamentals

Soil Mechanics Fundamentals

Editor

Marian Sewick

Soil Mechanics Fundamentals

Edited by **Marian Sewick**

ISBN: 978-1-68117-172-2
Library of Congress Control Number: 2015951996

Notice

Reasonable efforts have been made to publish reliable data and views articulated in the chapters are those of the individual contributors, and not necessarily those of the editors or publishers. Editors or publishers are not responsible for the accuracy of the information in the published chapters or consequences of their use. The publisher believes no responsibility for any damage or grievance to the persons or property arising out of the use of any materials, instructions, methods or thoughts in the book. The editors and the publisher have attempted to trace the copyright holders of all material reproduced in this publication and apologize to copyright holders if permission has not been obtained. If any copyright holder has not been acknowledged, please write to us so we may rectify.

Printed in United States of America on Acid Free Paper

Preface

Soil mechanics is the study of the physical properties of soil, especially those properties that affect its ability to bear weight, such as water content, density, strength, etc.

Soil mechanics is a branch of engineering mechanics that describes the behavior of soils. It differs from fluid mechanics and solid mechanics in the sense that soils consist of a heterogeneous mixture of fluids and particles but soil may also contain organic solids, liquids, and gasses and other matter. Along with rock mechanics, soil mechanics provides the theoretical basis for analysis in geotechnical engineering. Soil mechanics is used to analyze the deformations of and flow of fluids within natural and man-made structures that are supported on or made of soil, or structures that are buried in soils. Applications are building and bridge foundations, retaining walls, dams, and buried pipeline systems. Principles of soil mechanics are also used in related disciplines such as engineering geology, geophysical engineering, coastal engineering, agricultural engineering, hydrology and soil physics.

Soil Mechanics covers the genesis and composition of soil, the distinction between pore water pressure and inter-granular effective stress, capillary action of fluids in the pore spaces, soil classification, seepage and permeability, time dependent change of volume due to squeezing water out of tiny pore spaces, also known as consolidation, shear strength and stiffness of soils.

Table of Contents

CHAPTER 3   Experimental Study of Dynamic Characteristics on Composite Foundation with Cfg Long Pile and Rammed Cement-Soil Short Pile 53

CHAPTER 4   Influence of the Soil-Structure Interaction in The Behavior of Mat Foundation 73

CHAPTER 1

Improved Performance of Electrical Transmission Tower Structure Using Connected Foundation in Soft Ground

Doohyun Kyung [1], Youngho Choi [2], Sangseom Jeong [1] and Junhwan Lee [1]

[1]School of Civil and Environmental Engineering, Yonsei University, 134 Shinchon-dong, Seodaemun-gu, Seoul 120-749, Korea;
[2]KTP Consultants Pte Ltd., E-Centre@Redhill, 3791 Jalan Bukit Merah, Singapore 159471, Singapore

ABSTRACT

A connected foundation is an effective foundation type that can improve the structural performance of electrical transmission towers in soft ground as a resilient energy supply system with improved stability. In the present study, the performance of a connected foundation for transmission towers was investigated, focusing on the effect of connection beam properties and soil conditions. For this purpose, a finite element analysis was performed for various foundation and soil conditions. In order to validate the finite element analysis, the calculated results were compared with measured results obtained from field load tests. The use of connection beams was more effective for uplift foundations that usually control the design of transmission tower foundations. For the effect of soil condition, the use of connected foundation is more effective in soft clays with lower undrained shear strength (s_u). Smaller amounts of differential settlement were observed in all soil conditions for both unconnected and connected foundations when a bearing rock layer was present. When the foundation was not reinforced by connection beams, the values of lateral load capacity of tower structure (H_u) were similar for both with- and without-rock layers. It was confirmed that introducing haunch-shaped connection beams is effective for increasing connection beam stability.

KEYWORDS

Resilient energy infrastructure system; Electrical transmission tower structures; Connected foundation; Soft grounds; Field load test; Finite element analysis

INTRODUCTION

For the electrical transmission system, transmission towers are often installed with certain foundations that support the upper tower structure and associated overhead power lines. The transmission tower structures are subjected to various unexpected damages from a wide range of extreme weather events, including hurricanes, tornadoes, snow, and ice storms, and human disasters such as terrorism [1,2,3]. In particular, as climate change has been an issue in various social and engineering fields, it has become important to prepare a resilient infrastructure system that can guarantee or improve the stability of the energy supply system [4,5,6]. For the subsurface soil zone where the foundations of transmission towers are embedded, the issue of climate change also needs to be addressed for both design and construction of the structures. Increasing freezing-thawing cycles within the soil zone cause changes in various soil properties such as permeability, volume change behavior, strength, and compressibility [7,8,9]. Unusual fluctuation of groundwater level due to changes in annual precipitation characteristics causes additional settlements and unexpected reduction in the bearing capability of foundations [10,11,12]. All these threaten the stability and sustainability of the electrical transmission tower system, highlighting the need for a more robust and resilient structural system with certain reinforcements.

The types of tower foundation often used are pile, pier, inverted T-type, and mat foundation [13,14,15,16]. The inverted T-type foundation is widely used in transmission towers and can be used for small load conditions in good quality soils composed of sand. The pier foundation can be used in steep grade or deep bearing strata. This foundation is effective to support large loads of transmission towers and is frequently used for Ultra High Voltage (UHV) transmission towers. Pile foundation is generally used in weak soils, such as clay and reclaim soil, and often suffers structural damage and geotechnical instability due to insufficient foundation resistance and large differential settlements [14,15,17].

The size of transmission towers increases with electricity demand; the foundation size also increases to efficiently support the larger transmission towers. A foundation reinforced with additional structures is often used to improve foundation performance. For example, various researchers proposed setting rock bolt on a pier foundation to reduce settlement and to increase the resistance of the foundation; and setting a protective slab under the tower foundation to reduce the differential settlement of the foundation [18,19,20].

TEPCO [21] and IEEE [13] proposed the use of connection beams placed between the individual tower foundation components. According to TEPCO [21], the increase in load capacity for a foundation reinforced with a connection beam can be estimated based on the mobilized shear stresses and bending moments due to the weight of the connection beam. In the IEEE [13], a general description of the use of connection beams is presented with emphasis on the reduction effect of differential settlements. The significant effects of connection beams were investigated by Kyung *et al.* [22]. They performed small-scale model tests and finite element analyses in a clay condition and found that a 25% relative stiffness of the connection beams to that of the mat foundation is most effective to improve foundation performance.

All these results were obtained for certain assumed soil and foundation conditions. There were no changes in soil conditions such as strength and compressibility, which were indicating limited condition of the application. As a connected foundation can be more effectively used in soft, clayey soil, it is important to check any possible effect of soil and foundation conditions on variation in mechanical performance.

In the present study, the effects of a connected foundation on the performance of transmission tower structures were investigated with consideration of different types of connection beam and various soil conditions. Soft clay conditions with and without a bearing rock layer were assessed as well. For this purpose, a series of finite element analyses were performed and used to analyze the effects of a connected foundation. A large-scale field load test using a prototype model structure was performed and compared with the results from the finite element analyses. Improved performance of transmission tower structures was analyzed in detail for various connection beam stiffness and soil conditions.

TRANSMISSION TOWER FOUNDATION

Foundation Types and Design Procedure

A transmission tower structural system consists of overhead power lines, steel lattice tower structures, and foundations. The foundations are generally installed at the four corners of the tower structure. The foundations of transmission towers can be classified as axial load foundations and moment load foundations. Axial load foundations indicate the cases where lateral load acting on a tower would be transferred as uplift and compressive loads on the individual foundations at each corner. Pile foundations, pier foundations, and inverted T foundations, shown in Figure 1a to c, are considered as axial load foundations, which are effective in resisting lateral tower loads but vulnerable to differential settlements. A mat foundation, shown in Figure 1d, is a moment-load foundation, which is effective in preventing structural damage from differential settlements, while the lateral resistance tends to be lower than that of axial load foundations.

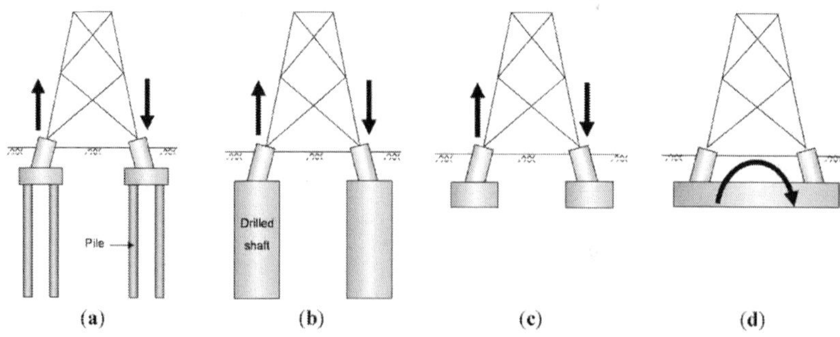

Figure 1. Types of transmission tower foundations: **(a)** pile foundation; **(b)** pier foundation; **(c)** inverted-T foundation; and **(d)** mat foundation.

The type of transmission tower foundation used is determined based on soil conditions. A pile foundation is generally used in soft soil conditions. For stability analysis, the load capacity of an individual foundation is evaluated for a given soil condition, ensuring that the load capacity is greater than the design load. Figure 2 shows the typical configuration of loads and resistances for a transmission tower structure where an applied lateral load on a tower (H), transferred loads to lower foundations (Q), and mobilized foundation resistances (R) are indicated. The lateral load (H) acting on a tower is transferred to the lower foundations of the front compressive and rear uplift sides. The transferred loads are composed of the vertical (Q_{vc} and Q_{vt}) and horizontal (Q_{hc} and Q_{ht}) load components, as indicated inFigure 2. The magnitude of each load component can be calculated from the geometric characteristics of the transmission tower. The stability of a transmission tower foundation is then determined as follows [21,23]:

$$Q_{vc} \leq R_{vc,m} = \frac{R_{vc}}{FS} \text{ and } Q_{vt} \leq R_{vt,m} = \frac{R_{vt}}{FS} \tag{1}$$

$$Q_{hc} \leq R_{hc,m} = \frac{R_{hc}}{FS} \text{ and } Q_{ht} \leq R_{ht,m} = \frac{R_{ht}}{FS}, \tag{2}$$

where Q_{vc} and Q_{vt} = transferred compressive and uplift tensile loads on the front and rear sides; Q_{hc} andQ_{ht} = transferred horizontal loads on the front and rear sides; $R_{vc,m}$ and $R_{vt,m}$ = allowable compressive and uplift resistances; $R_{hc,m}$ and $R_{ht,m}$ = allowable horizontal front and rear resistances; R_{vc}, R_{vt}, R_{hc}, andR_{ht} = ultimate compressive, uplift, and horizontal front and rear resistances; and FS = factor of safety. While stabilities for both vertical and horizontal loads must be guaranteed, the vertical stability against uplift load (Q_{vt}) frequently controls the design, as uplift resistance is usually smaller than vertical compressive resistance.

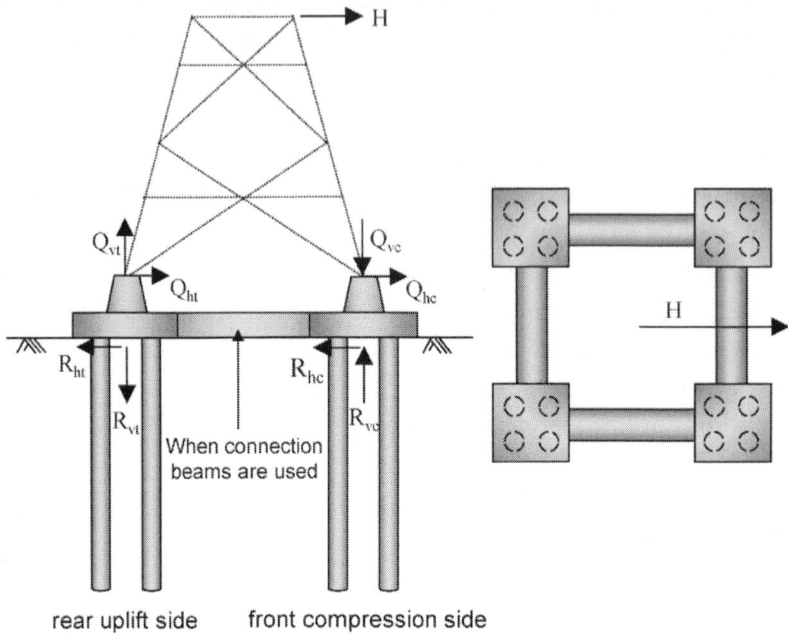

Figure 2. Configurations of loads and resistances for transmission tower structure.

Connected Foundation

When lateral loads act on the upper transmission tower, the lower foundations at the corners are subject to uplift and compressive loads. This indicates that a larger amount of differential settlement could occur due to the opposite directions of induced settlements at the uplift and compressive sides, which may cause structural damage to the entire transmission tower structure system. Modified foundations reinforced using additional structural components are an option to reduce differential settlement and improve the performance of a foundation. For example, Yang *et al.* [24] and Wang *et al.* [25] presented connected H-shaped girders to prevent instability in transmission towers. Yuan *et al.* [20] also presented tower foundations with protective slabs in order to reduce differential settlement.

A connected foundation is a type of reinforced foundation using connection beams placed between foundations, as illustrated in Figure 2.

The design guidelines and performance analysis of connected foundations can be found in TEPCO [21], IEEE [13], and Kyung *et al.* [22]. According to TEPCO [21], connection beams are regarded as rigid components; their mechanical properties are not considered in the design. IEEE [13] also referred to the use of connection beams for the same purpose of increasing foundation resistance and reducing differential settlements. Kyung *et al.* [22] analyzed the performance of connected foundations for different structural and load conditions. It should be noted that all these investigations were performed on the limited soil and foundation conditions available in the experimental testing program. Further investigation is warranted to examine the detailed performance of connected foundations in different foundation and soil conditions.

FINITE ELEMENT ANALYSIS

Description of Analysis

Full-scale model tests of foundations for transmission towers are difficult due to the great expense and effort required in constructing test models and conducting field load tests. For this reason, numerical modeling and analysis are often introduced as an alternative. In the present study, the finite element (FE) analysis of foundations for full-scale transmission tower structures was performed to analyze the behavior and improved performance of foundations reinforced with connection beams. The PLAXIS 3D Foundation [26] was used in the study, which is a widely used finite element analysis program for geotechnical engineering.

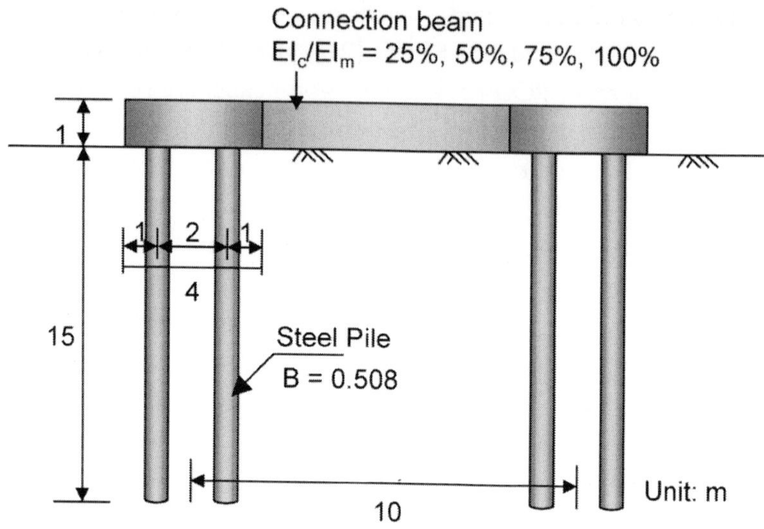

Figure 3. Configuration of foundation parts for finite element analysis.

The finite element analyses were planned and prepared based on the typical size and configuration of 345 kV transmission tower structures. Figure 3 shows the configuration of transmission tower foundation considered in the finite element analyses. As shown in Figure 3, the foundation parts are placed at the four sides of the transmission tower, and the foundation at each side consists of a mat and four piles. The width and height of the mat are 4 and 1 m, respectively, and the distance between the foundation sides is 10 m. The diameter and length of the piles are 0.508 and 15 m, respectively, and the pile-to-pile spacing equals 2.7 m.

Four different sizes of connection beams were considered in the analyses, as the rigidity of connection beam varies with its size and affects the effectiveness of connection beams. While the height of the connection beams was always 1 m, different widths of connection beams equal to 25%, 50%, 75%, and 100% of the mat width were considered. The material properties of the connection beams were the same as those of the mat with an elastic modulus (E) and Poisson's ratio (υ) of 40 GPa and 0.2, respectively. This means that the flexural stiffness of the connection beams (EI) was equal to 25%, 50%, 75%, and

100% of the mat EI. The height of the tower structure was 27 m, and the transferred loads on the foundations were determined from the geometric condition of the transmission tower and foundations when the lateral load (H) was applied on the top of the tower structure.

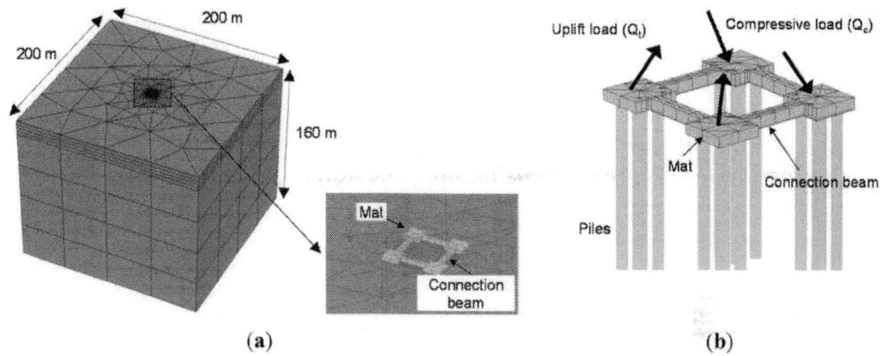

Figure 4. Finite element model for transmission tower foundation: **(a)** configuration of connected foundation model and **(b)** configuration of applied loads.

Figure 4 shows the finite element model prepared in this study. The width and depth of the finite element model in this study were 200 and 160 m, respectively, corresponding to a size larger than 10 times the pile length or 23 times the mat width. According to Liang *et al.* [27], this represents a sufficiently larger model size to avoid the boundary effect. Finer mesh was used in the zone near the model structure where higher stress concentration was expected. Fixed-end boundary conditions were set along the lateral and bottom sides of the mesh. Mat foundations were assumed to be linear-elastic concrete material, as described previously. Piles were regarded as linear-elastic steel piles with an elastic modulus (E) and Poisson's ratio (υ) equal to 205 GPa and 0.3, respectively. The interface elements were used between the soil and the pile surfaces, which allowed elastic small displacements and plastic slip behaviors. The shear strength of the interface element was defined based on the strength reduction factor R_{int} [26], which is given by the following relationships:

$$c_{\text{int}} = R_{\text{int}} c_{soil}$$
(3)

$$\tan \phi_{\text{int}} = R_{\text{int}} \tan \phi_{soil}. \qquad (4)$$

where c_{int} and ϕ_{int} = cohesion and friction angle of interface; and c_{soil} and ϕ_{soil} = cohesion and friction angle of soil materials; and R_{int} = reduction factor. The value of R_{int} was set to 0.4, as obtained from the model load test results that will be further described in Section 3.3. Figure 4b shows the detailed configuration of the applied loads on the foundations at corners. With lateral load acting on the tower, the two front-side foundations are subjected to compressive loads and the other two rear-side foundations are subjected to uplift loads.

Soil Conditions
Various clay soil conditions were considered in the finite element analyses. Figure 5 shows the soil profiles considered in the finite element analyses. The soil conditions considered were a soft clay deposit with a lightly over-consolidated (OC) layer near the surface, as is commonly encountered in practice. Below the surface layer, soils were all assumed to be normally consolidated (NC) clay. Other cases with a bearing rock layer at the pile base level, as indicated in Figure 5a, were also considered in the finite element analyses. The unit weight (γ_{sat}) of clay was equal to 16 kN/m^3 with a Poisson's ratio (υ) of 0.495 assuming undrained condition. For the NC clay condition, the undrained shear strength (s_u) varied linearly with depth given as follows:

$$s_u = \alpha \cdot \gamma' \cdot z \qquad (5)$$

where α = strength increase ratio; z = depth; and γ' = effective unit weight. The values of s_u within the OC zone near the surface were assumed to be 10 kPa. α is the ratio of s_u to the vertical effective stress (*i.e.*, $\gamma' \cdot z$). The value of α for most clays ranges from 0.2 to 0.4 [28,29,30,31,32]. In the present study, the value of α was set to 0.2, assuming a soft clay condition.

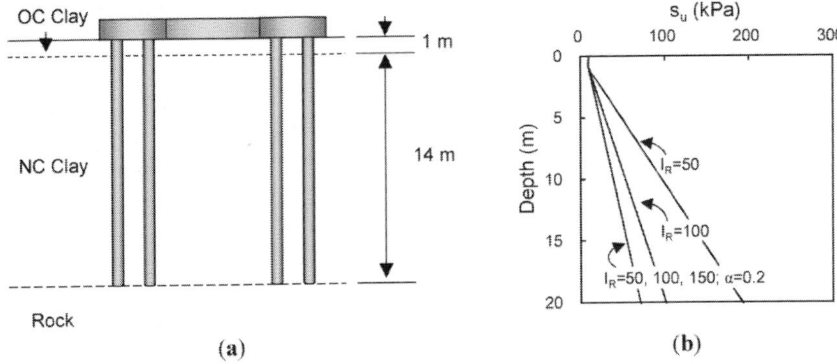

Figure 5. Soil conditions for finite element analysis: **(a)** soil layer condition and **(b)** depth profile of considered soil condition.

The rigidity index is often used to characterize the mechanical state of clay and is defined as the ratio of shear modulus (G) to s_u given by:

$$I_R = \frac{G}{s_u} = \frac{E_s}{2 \cdot (1 + \nu_u) \cdot s_u} = \frac{E_s}{3 \cdot s_u} \tag{6}$$

where G = shear modulus; E_s = elastic modulus; and υ_u = Poisson's ratio for undrained condition = 0.5. Three different values of I_R equal to 50, 100, and 150 were considered in the analyses. Figure 5b shows the profiles of soil conditions considered in the finite element analyses. For the cases with a bearing rock layer, the values of E and υ for rock were 50 MPa and 0.15, respectively, which represent the conditions of weathered rock [33,34].

Comparison of Finite Element Analyses with Field Load Tests

To assess the validity of the finite element analysis, results from field load tests for the transmission tower structures were obtained and compared with those from the finite element analyses. Two field test cases were used, the small-scale model load tests reported in Kyung *et al.* [22] and larger-scale prototype model load tests conducted in the present study. For both cases, the transmission tower structures were prepared using connected foundations.

The small-scale model transmission structures consisted of an upper tower structure and lower foundations, as shown in Figure 6a. The foundation at each corner included a mat and piles that were embedded into the ground to a depth of 0.8 m. The width and height of the mat were 0.1 and 0.05 m and the diameter and length of the piles were 0.05 and 0.8 m, respectively. Two different types of connection beams were used in the tests: (1) low-stiffness beams with EI = 6.135 N·m² and (2) high-stiffness beams with EI = 1571 N·m².

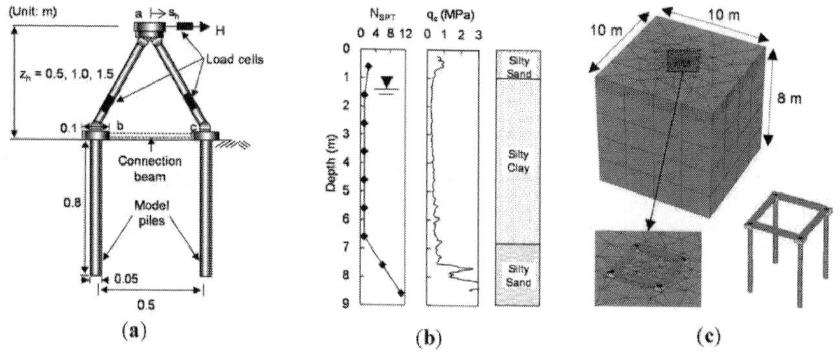

Figure 6. Small-scale model load tests with connected foundations [22]: (a) model structures; (b) depth profiles of soil layer, SPT, and CPT results; and (c) FE model configuration.

The soils at the test site were clays classified into CL according to the unified soil classification system (USCS). The unit weight (γ_{sat}), specific gravity (G_s), water content (w), and plasticity index (PI) were 16.59 kN/m³, 2.69, 43.3%, and 23.3%, respectively. Unconfined compression and unconsolidated undrained (UU) triaxial tests were conducted, and s_u ranged from 8.4 to 11.1 kPa. Figure 6b shows the detailed soil profiles of the standard penetration (SPT) and cone penetration (CPT) test results. The detailed configuration of the finite element model for the small-scale model load tests is shown in Figure 6c. As the clay was quite homogeneous and pile length was only 0.8 m, a single layer was assumed in the finite element analysis with material properties given in Table 1. Mat and pile were assumed as linear-elastic materials with the elastic modulus (E) and Poisson's ratio (υ) equal to 205 GPa and 0.3, respectively. The Mohr–Coulomb model was adopted with the undrained shear strength given in

Table 1.Figure 7 shows the compared uplift load-displacement curves from the small-scale model tests and finite element analyses for two connection beam cases of EI = 6.135 and 1571 N·m². As shown in Figure 7, reasonably close agreements are observed between measured and estimated results.

Table 1. Material properties for small-scale model load test.

Material	Model	Depth (m)	γ_t^1 (kN/m³)	E^2 (kN/m²)	s_u^3 (kN/m²)	υ^4
Clay	Mohr–Coulomb	0–8	16.59	3,000	11.08	0.5
Mat	Linear elastic	-	75	205,000,000	-	0.3
Pile	Linear elastic	-	75	205,000,000	-	0.3

γ_t^1 = total unit weight, E^2 = elastic modulus, s_u^3 = undrained shear strength, υ^4 = Poisson's ratio.

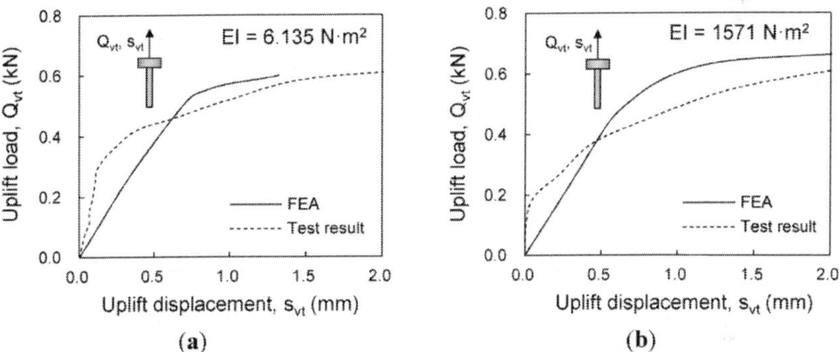

Figure 7. Compared load-displacement curves of small-scale model tests and finite element analyses for connection beam cases of **(a)** EI = 6.135 N·m² and **(b)** EI = 1571 N·m².

The other field load test using a larger-scale prototype model was conducted in this study at Hwaseong, Korea. Figure 8a shows the configuration of the prototype model structure. The load height (z_h) and contiguous length (w) of the transmission tower structures were 15 and 4.84 m, respectively. The mat foundations were made of concrete with width and height equal to 2.4 and 0.8 m, respectively. Four piles with a diameter of 0.318 m were used at the corners and were embedded to the depth of 17 m. The connection beams with 25% mat stiffness were used for this prototype structure. The width, height, and length of the connection beams were 0.6, 0.8, and 2.44 m, respectively.

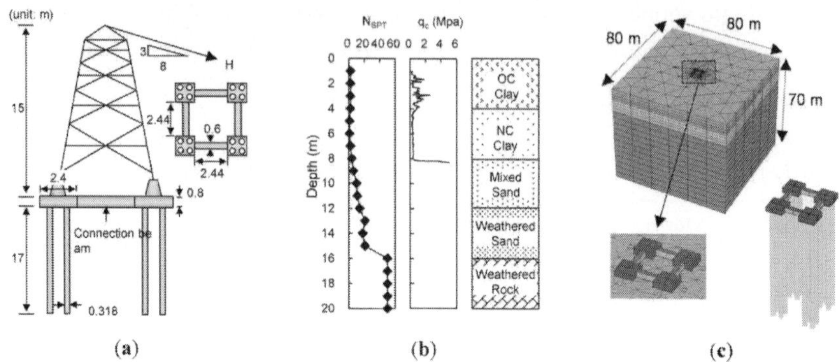

Figure 8. The prototype model test for connected foundation: **(a)** detailed configuration of prototype model; **(b)** site investigation results for the site; and **(c)** FE model configuration.

Figure 8b shows the depth profiles of soil layers, SPT, and CPT results at the test site. It is seen that the top 4 m of soil was a slight OC clay layer, below which an NC clay layer extended down to the depth of 8 m. Mixed and weathered sand layers were observed to depths of 13 to 16 m, beneath which was a weathered rock layer. The pile base was placed on this weathered rock layer. The layered soil condition was considered for modeling the prototype model load test in the finite element analysis. Figure 8c shows the configuration of finite element model for the prototype model load test and Table 2 shows the material properties used in the analysis.

Table 2. Material properties for the prototype model load test.

Material	Model	Depth (m)	γ_t^1 (kN/m^3)	E^2 (kN/m^2)	s_u^3 (kN/m^2)	ϕ'^4 (°)	υ^5
OC Clay	Mohr–Coulomb	0–4	18.7	2900	5–64	-	0.5
NC Clay	Mohr–Coulomb	4–8	18.7	2900	20	-	0.5
Mixed Sand	Mohr–Coulomb	8–12	19.5	2,200–5,200	-	27	0.3
Weathered Sand	Mohr–Coulomb	12–16	20.5	5,200–12,000	-	30	0.35
Weathered Rock	LE	16–70	25	300,000	-	-	0.25
Mat	LE	-	25	40,000,000	-	-	0.2
Pile	LE	-	75	205,000,000	-	-	0.3

γ_t^1 = total unit weight, E^2 = elastic modulus, s_u^3 = undrained shear strength, ϕ'^4 = friction angle, υ^5 = Poisson's ratio.

Figure 9 compares the results from the prototype model load test (PMT) and finite element analysis (FEA). As shown in Figure 9a,b, the compressive and uplift load responses of the connected foundations from the finite element analyses were in close agreement with the measured results from the field load test. Figure 9c shows the estimated and measured vertical displacement profiles at lateral load of H = 1200 kN. The profiles of vertical and differential settlements also show reasonable agreement between FEM and PMT results.

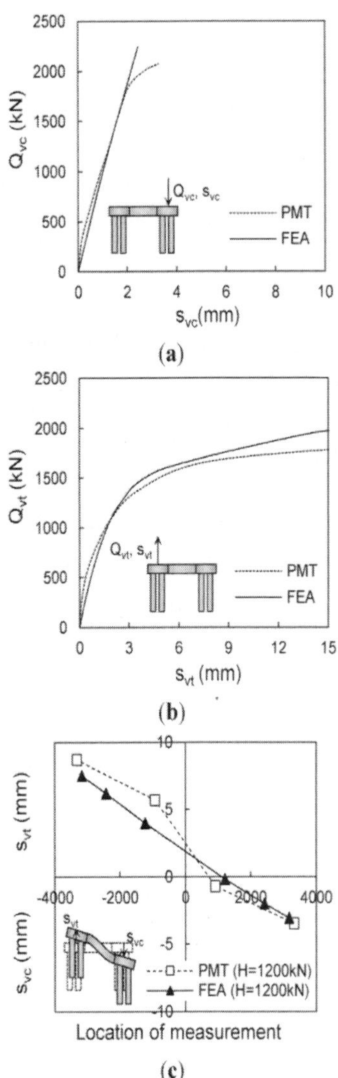

Figure 9. Compared load-displacement curves of prototype model test (PMT) and finite element analysis (FEM): **(a)** compressive load responses; **(b)** uplift load responses; and **(c)** vertical settlement profiles.

PARAMETRIC STUDY ANALYSIS

Effects of Connection Beam Stiffness

Figure 10 shows the load-displacement curves obtained from the finite element analyses for unconnected and connected foundations. The results for different ratios of connection-beam stiffness (EI_c) to mat stiffness (EI_m) were all included in Figure 10. In Figure 10a,b, the uplift and compressive load-displacement curves were plotted for clay with I_R = 100. As shown in Figure 10, both the uplift and compressive load-carrying capacities increase as the connection beam stiffness increases. It is also noted that the amount of load-carrying capacity increase from the connected foundation was higher for the uplift foundation case than for the compressive foundation components. This is due to a higher confining effect on the uplift side where larger displacements and lower load carrying capacity are observed. When the uplift and compressive sides are connected, however, uplift and compressive displacements become similar due to the effect of the connection beam. This means that the use of connection beams would be more effective for the uplift foundations that usually control the design of transmission tower foundations.

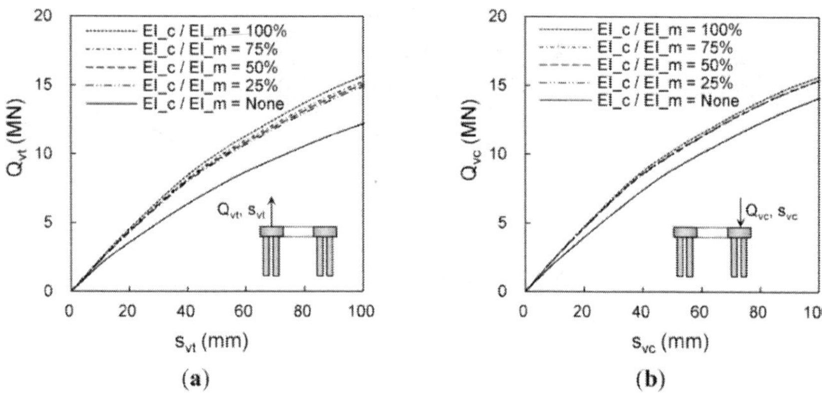

(a)   (b)

Figure 10. Load-displacement curves with different connection beam of conditions: (**a**) uplift foundation and (**b**) compressive foundation components.

Figure 11 shows the reductions in vertical displacement with relative connection-beam stiffness (EI_c/EI_m). All displacements in Figure 11 were measured at a load equal to the ultimate load capacity of the unconnected foundation. Figure 11a,b shows those for the uplift displacement (s_{vt}) and compressive settlement (s_{vc}), respectively. The values of s_{vt} and s_{vc} both decreased with increasing connection beam stiffness. The reduction effect increased with increasing EI_c/EI_m but the reductions were relatively small after EI_c/EI_m equal to 25%. The reductions in uplift displacement (s_{vt}) in Figure 11a were larger than those of compressive settlement (s_{vc}) in Figure 11b. Reductions in differential settlements for unconnected and connected foundations are shown in Figure 11c. Note that the variation of differential settlement occurrence with EI_c/EI_m follows the same tendency to those of tensile and compressive displacements. Approximately 21–25% reductions were observed, showing similar tendency and efficiency to those shown in Figure 11a,b.

Figure 12 shows the lateral load capacity H_u of tower structure and uplift and compressive load capacities of connected foundation for different EI_c/EI_m. As shown in Figure 12a, increases in H_u show a similar tendency to those in Figure 11. H_u increased by 25%, 26%, 28%, and 32% for EI_c/EI_m values of 25%, 50%, 75%, and 100%, respectively.

Figure 12b shows the transferred uplift and compressive loads of $Q_{vt,p}$ and $Q_{vc,p}$ on the piles at the rear uplift and front compressive sides, respectively. Note that $Q_{vc,p}$ and $Q_{vt,p}$ do not include the mat capacity and thus $Q_{vc,p}$ is different Q_{vc}. Both $Q_{vc,p}$ and $Q_{vt,p}$ decreased with increasing connection beam stiffness due to the load sharing effect of connection beams. For the unconnected foundation ($EI_c/EI_m = 0$), $Q_{vt,p}$was larger than $Q_{vc,p}$ due to the contact resistance of the mat on the ground at compressive side. For the connected foundation, however, the stiffer the connection beam with increasing EI_c/EI_m, the smaller the difference between $Q_{vc,p}$ and $Q_{vt,p}$. This is because connection beams confine the uplift and compressive foundations, together producing combined behavior.

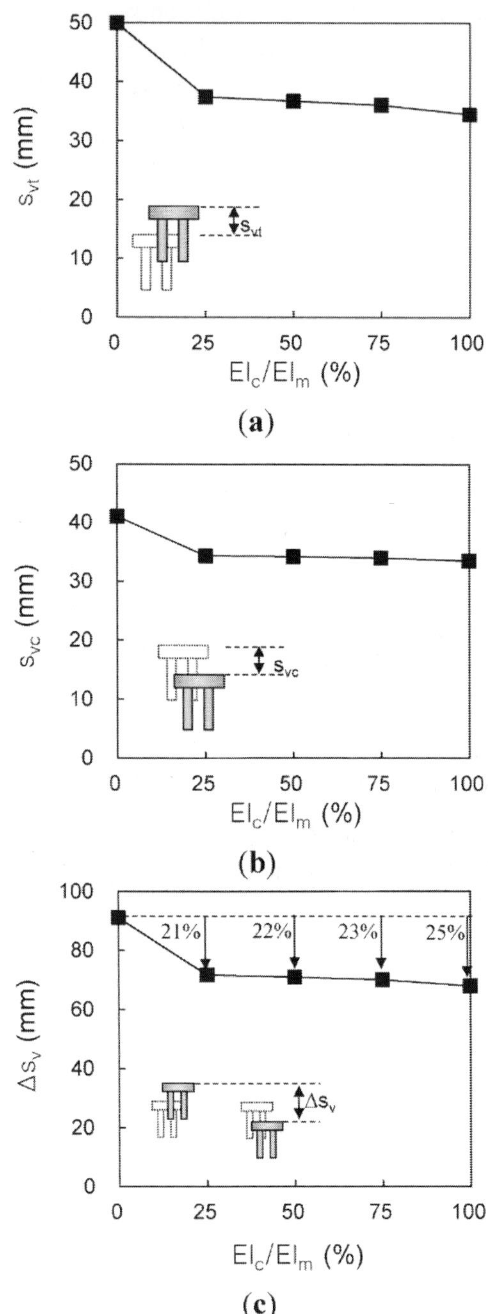

Figure 11. Reductions in vertical displacements with connection beam: (**a**) uplift displacement (s_{vt}); (**b**) compressive settlement (s_{vc}); and (**c**) differential settlement (Δs_v).

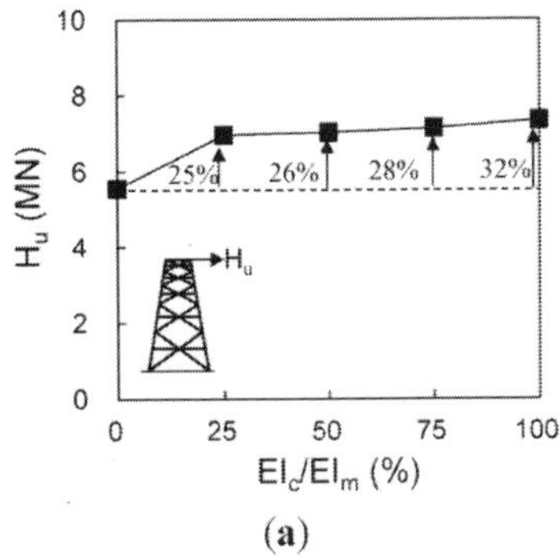
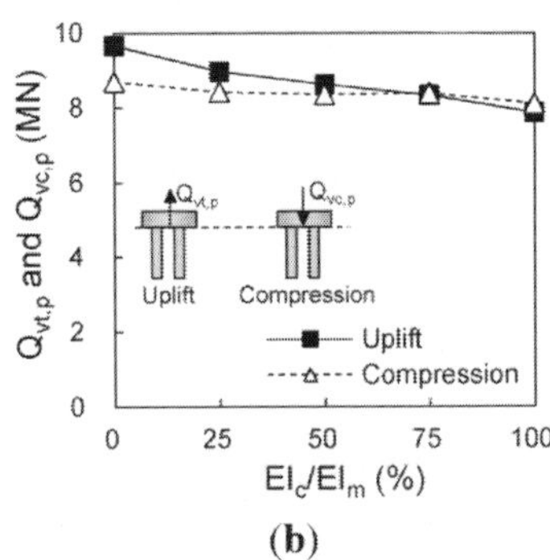

Figure 12. Load capacities of tower structure and connected foundation: **(a)** ultimate lateral load capacity of transmission tower (H_u) and **(b)** transferred uplift ($Q_{vt,p}$) and compressive ($Q_{vc,p}$) loads.

Effects of Soil Condition

To check the effect of soil condition on the performance of connected foundation, changes in differential settlement and H_u from connected foundations were obtained from the results of finite element analysis and plotted in Figure 13a,b with I_R, respectively. As I_R represents the combined characteristics of s_u and E_s, for which values increase with soil depth (z), cases with different values of s_u and E_s were considered and included in Figure 13. All results in Figure 13 were obtained for the 25% relative connection beam stiffness. As shown in Figure 13a, larger reductions in differential settlement are observed as I_R increases for both increasing E_s and decreasing su. For $I_R = 100$, the higher E_s case showed higher reduction. The results in Figure 13 indicate that reductions in differential settlement become larger as soil becomes stiffer with higher I_R condition. This is because stiffer clays with higher I_R produce less compressive settlement while changes in uplift displacements were smaller.

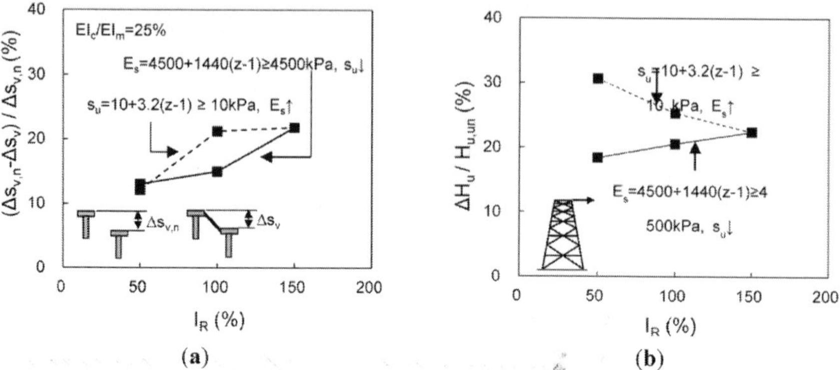

Figure 13. Effects of soil conditions on connected foundation: (**a**) reduction ratio of differential settlement and (**b**) increases in lateral load capacity of transmission tower.

Figure 13b shows the ratios of increase in lateral load capacity (ΔH_u) for the transmission tower with connected foundation to H_u for unconnected case. It is seen that $\Delta H_u/H_u$ decreases with increasing E_s and increases with decreasing s_u. These results indicate that $\Delta H_u/H_u$ increases more in weaker soil conditions with lower s_u than for higher E_s cases. From the results in Figure 13a,b, it can be summarized that the use of a connected foundation is more effective in soft clays with lower s_u. For the effect of soil stiffness, however, different tendencies were observed for differential settlement and lateral load capacity.

Effect of Bearing Rock Layer

Figure 14 shows the compressive and uplift load-displacement curves of connected foundations with and without a bearing rock layer. The compressive load-displacement curves for the with-rock layer condition are much higher than those for the without-rock layer condition. This can be attributed to increases in the end-bearing capacity of the piles that extend to the depth of the bearing rock layer. However, the uplift load-displacement curve demonstrates no major differences between with or without-rock layer conditions. This is because the existence of a rock layer does not contribute much to the uplift pile capacity that is mobilized on the pile shaft.

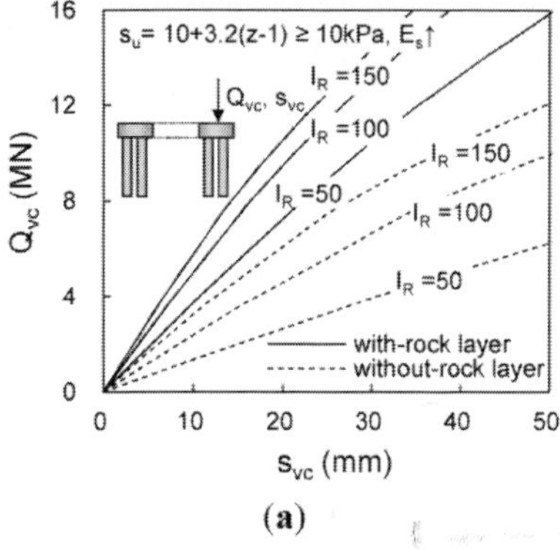

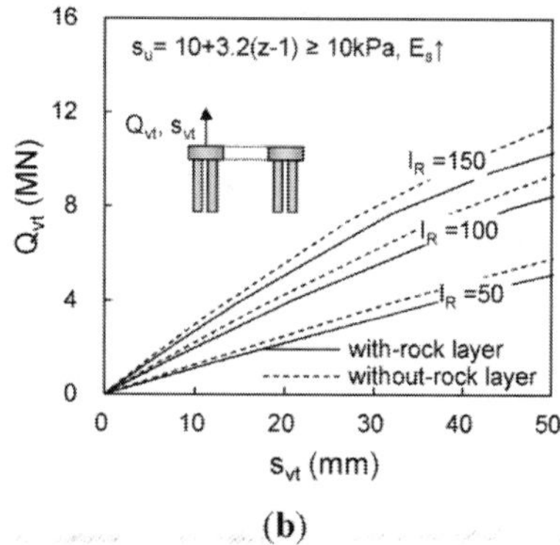

Figure 14. Load-displacement curves of connected foundations with and without bearing rock layer: (**a**) compressive and (**b**) uplift foundation.

Figure 15 shows the changes in differential settlement and lateral load capacity with and without a bearing rock layer. As shown in Figure 15a, smaller amounts of differential settlement were observed in all soil conditions for both unconnected and connected foundations when a bearing rock layer was present. This is due to restricted settlement of the pile base that is resting on the rock layer. It is also seen that reductions in differential settlement with the presence of connection beams were greater for the without-rock layer condition than for the with-rock layer condition.

Figure 15b shows that the lateral load capacity (H_u) of the transmission tower was higher in the absence of a rock layer than in the presence of a rock layer. When the foundation was not reinforced by connection beams, the values of H_u were similar for both with- and without-rock layer because each foundation behaves independently and the uplift capacity controls H_u anyway. When the foundation was reinforced by connection beams in a rock layer, the uplift displacement experienced a greater increase than the one that occurs in the absence of a rock layer because the overturning axis was closer to the compressive foundation, which was caused by decreasing compressive displacement. Therefore, the rate of H_u increase for connected foundations becomes lower when a bearing rock layer exists.

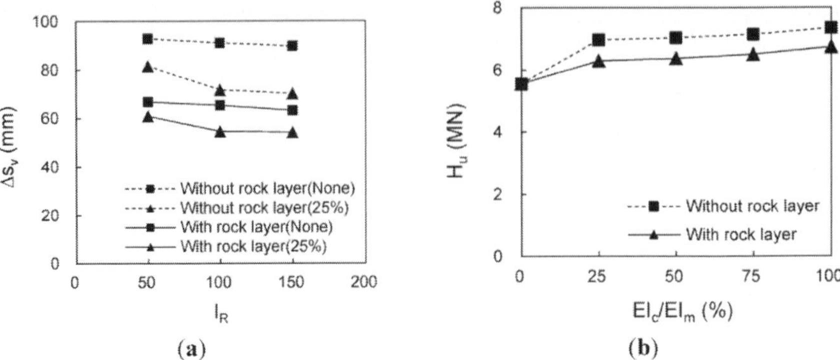

Figure 15. Differential settlement and lateral load capacity for different rock layer conditions: (**a**) differential settlement and (**b**) lateral load capacity of tower structure.

Bending Moment Distribution for Connection Beam

The connected foundation of the transmission tower is subjected to uplift and compressive loads on different sides, and bending moment occurs in the connection beam. The bending moment in the connection beam can cause structural damage, while this can be prevented or reduced by changing the shape of the beam. Figure 16a shows the distribution of bending moment along the connection beam in different soil conditions. As shown in Figure 16a, the bending moment was largest at ends and decreased with distance from the ends. It is seen that the distribution of bending moment is not symmetrical and shows the location of zero bending moment at the distance of 1/4 length from the uplift-side of the connection beam. Minor differences in the bending moment were observed for the results with different soil conditions.

In order to decrease the amount of bending moments of the connection beam, haunch-shaped connection beams were considered and analyzed in the finite element analysis. As shown in Figure 16b, the haunch shapes were defined by the ratio of D_m to D_c. The value of D_c was set equal to 0.61 m corresponding to 1/4 of connection beam length. The values of D_m were then changed to consider different haunch shapes. Three haunch-shaped connection beams were used in the finite element analysis. The considered D_m/D_c values were equal to 2/5, 3/5, and 4/5. Figure 16b shows the distribution of bending moments for the considered haunch shapes. The magnitude of bending moment of the connection beams decreased with increasing D_m/D_c showing 6.1, 4.7, and 4.1 kN·m for uplift side and 5.1, 4.1, and 3.4 kN·m for the compressive side with D_m/D_c equal to 2/5, 3/5, and 4/5, respectively. This confirms that introducing haunch-shaped connection beams is effective for increasing connection beam stability.

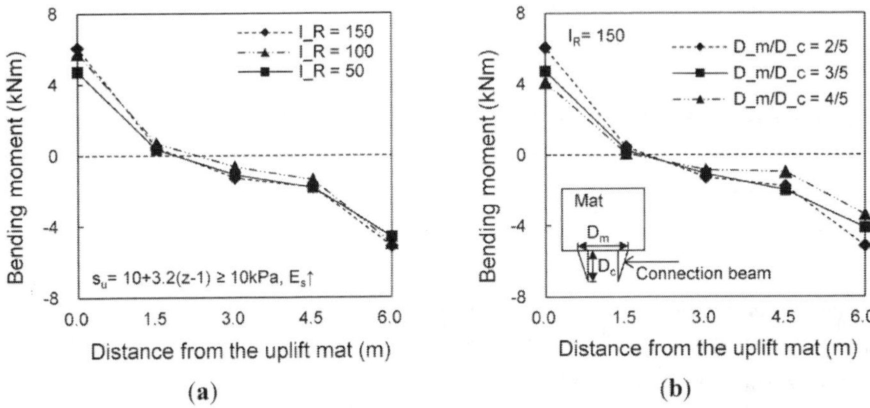

Figure 16. Distribution of bending moment along connection beams: **(a)** bending moment distribution for different soil conditions and **(b)** bending moment distributions for different haunch shapes.

SUMMARY AND CONCLUSIONS

In the present study, the effects of connected foundation on the performance of electrical transmission tower structures embedded in soft ground were investigated with consideration of various connection-beam and soil conditions. For this purpose, a series of finite element analyses were performed and used to analyze the improved performance of transmission tower foundation. To assess the validity of the finite element analysis, the field load test results were compared with those from the finite element analyses.

The application of connected foundation produced increases in resistance for uplift side higher than for compressive side. This indicates that the use of connection beams would be more effective for uplift foundations that usually control the design of transmission tower foundations. Uplift displacement and settlement both decreased with increasing connection beam, stiffness showing the most effective reduction ratio at EI_c/EI_m equal to 25%. The reductions in uplift displacement were larger than in compressive settlement. The lateral load capacity of a tower structure increased similarly to other resistance components.

For the effect of soil condition, it was seen that increases in the lateral load capacity H_u of a tower structure decreases with increasing E_s and increases with decreasing s_u. This implies that the use of a connected foundation is more effective in soft clays with lower s_u. The compressive load carrying capacity for the with-rock layer condition was much higher than for the without-rock layer condition due to an increase in the end-bearing capacity of piles. Smaller amounts of differential settlement were observed in all soil conditions for both unconnected and connected foundations when a bearing rock layer was present. When the foundation was not reinforced by connection beams, the values of H_u were similar for both with- and without-rock layers. It was confirmed that introducing haunch-shaped connection beams is effective for increasing connection beam stability.

The results obtained from this study indicate that the connected foundation is an effective option to improve the resilience of electrical transmission tower infrastructures with the effect of increasing load carrying capacity and reducing differential settlements. It is particularly effective within soft grounds where extra safety margins are necessary as there is increasing possibility of damage with weaker soil conditions. In addition, extreme weather events due to climate change can cause harmful changes in subsoil condition with chances of additional settlements and unexpected reduction in the bearing capacity of foundations. The design strategy for optimizing connected foundations of a transmission tower system given in this study can be used to prepare more robust and sustainable transmission infrastructures with improved resilience against various climate change scenarios and unfavorable soil conditions.

ACKNOWLEDGMENTS

This work was supported by a Power Generation & Electricity Delivery of the Korea Institute of Energy Technology Evaluation and Planning (KETEP) grant funded by the Korean Ministry of Knowledge Economy (No. 20101020200060). This work was also supported by the Basic Science Research Program through the National Research Foundation of Korea (NRF) grant funded by the Korean government (MSIP) (Nos. 2011-0030040 and 2013R1A1A2058863).

AUTHOR CONTRIBUTIONS

All three authors significantly contributed to the scientific study and writing. Doohyun Kyung and Junhwan Lee contribute to the overall idea, planning, financing, analyzing, and writing of the manuscript; Youngho Choi performed finite element analyses of connected foundations.

CONFLICTS OF INTEREST

The authors declare no conflict of interest.

REFERENCES

1. Ebinger, J.; Walter, V. *Climate Impacts on Energy Systems: Key Issues for Energy Sector Adaptation*; World Bank: Washington, DC, USA, 2011.
2. Yates, D.; Luna, B.Q.; Rasmussen, R.; Bratcher, D.; Garre, L.; Chen, F.; Tewari, M.; Hansen, P.F. Assessing climate change hazards to electric power infrastructure: A sandy case study. *IEEE Power Energy Mag.* 2014, *12*, 66–75.

3. Wilbanks, T.; Fernandez, S.; Backus, G.; Garcia, P.; Jonietz, K.; Kirshen, P.; Savonis, M.; Solecki, B.; Toole, L.; Allen, M.; *et al. Climate Change and Infrastructure, Urban Systems, and Vulnerabilities*; Technical report for the U.S. Department of Energy in Support of the National Climate Assessment, U.S. Department of Energy: Oak Ridge, TN, USA, 2012.

4. European Commission. Commission Staff Working Document: Adapting Infrastructure to Climate Change. In *An EU Strategy on Adaptation to Climate Change*; European Commission: Brussels, Belgium, 2013.

5. HM Government. Climate Resilient Infrastructure: Preparing for a Changing Climate. UK for The Stationery Office Limited on behalf of the Controller of Her Majesty's Stationery Office: London, UK; May; 2011.

6. Neumann, J.E.; Price, J.C. *Adapting to Climate Change: The Public Policy Response–Public Infrastructure*; An Initiative of the Climate Policy Program at RFF, Resources for the Future: Washington, DC, USA; June; 2009.

7. Konrad, J.; Samson, M. Hydraulic conductivity of kaolinite-silt mixtures subjected to close-system freeze and thaw consolidation. *Can. Geotech. J.* 2000, *37*, 857–869.

8. Othman, M.A.; Besnson, C.H. *Effect of Freeze-Thaw on the Hydraulic Conductivity of Three Compacted Clays from Wisconsin*; Transportation Research Board: Washington, DC, USA, 1992; pp. 118–125.

9. Yarbasi, N.; Kalkan, E.; Akbulut, S. Modification of the geotechnical properties, as influenced by freeze-thaw, of granular soils with waste additives. *Cold Reg. Sci. Technol.* 2007, *48*, 44–55.

10. Ausilio, E.; Conte, E. Influence of groundwater on the bearing capacity of shallow foundations.*Can. Geotech. J.* 2005, *42*, 663–672.

11. Shahriar, M.A.; Sivakugan, N.; Das, B.M.; Urquhart, B.; Tapiolas, M. Water table correction factors for settlement of shallow foundations in granular soils. *Int. J. Geomech. ASCE* 2015, *15*, 06014015.

12. Yasuhara, K.; Murakami, S.; Mimura, N.; Komine, H.; Recio, J. Influence of global warming on coastal infrastructural instability. *Sustain. Sci.* 2007, *2*, 13–25.

13. IEEE. IEEE Guide for Transmission Structure Foundation Design and Testing, IEEE Standard 691–2001. In Proceedings of the IEEE Power Engineering Society and the American Society of Civil Engineers, New York, NY, USA, 2001.

14. Jang, S.H.; Kim, H.K.; Ham, B.W.; Chung, K.S. A study on the transmission tower foundation design and construction method-A focus of cylindrical foundation. *J. Korean Inst. Electr. Eng. (KIEE)* 2007, *56*, 1031–1034.

15. KECA. *Handbook for Transmission Structure*; Korea Electrical Contractors Association: Seoul, Korea, 2003.

16. Morinaga, Y.; Kamiji, M.; Imoto, S.; Ogawa, S.; Iwamori, K. Transmission Tower Foundation in Japan. In Proceedings of the Transmission and Distribution Conference and Exhibition, Yokohama, Japan, 6–10 October 2002; Volume 3, pp. 2162–6165.

17. Kim, J.B.; Cho, S.B. The design and the full load test results of 765 kV tower foundation. In Proceeding of the Korean Institute of Electrical Engineers (KIEE) Fall National Conference, Seoul, Korea, November 1995; pp. 447–449.

18. Jeoung, S.S.; Ham, H.K.; Lee, D.S. Load transfer analysis of drilled shaft reinforced by soil nails. *J. Korean Geotech. Soc.* 2004, *20*, 37–47.

19. Nam, D.S.; Kim, S.I.; Lee, J.H.; Yoon, K.S. Optimization of Reinforcement Effect of Large-diameter drilled deep foundation. *J. Korean Geotech. Soc.* 2003, *19*, 207–216.

20. Yuan, G.L.; Li, S.M.; Xu, G.A.; Si, W.; Zhang, Y.F.; Shu, Q.J. The anti-deformation performance of composite foundation of transmission tower in mining subsidence area. *Prodedia Earth Planet. Sci.* 2009, *1*, 571–576.

21. TEPCO. *Design Guideline for UHV Foundation*; Tokyo Electric Power Company: Tokyo, Japan, 1988.

22. Kyung, D.H.; Kim, D.H.; Lee, J.H. Improved mechanical performance of connected foundations for transmission tower structures in soft soils. *Eng. Struct.* 2014. in submit.

23. KEPCO. *Design Standard for Transmission Tower Foundation*; DS-1110. Korea Electronic Power Corporation: Seoul, Korea; December; 2011.

24. Yang, J.S.; Yang, Y.H.; Yan, L.; Zhang, H.L.; Hu, X.; Tang, P. Construction scheme choice of large-span tunnels under-passing high voltage transmission tower and its application. *Chin. J. Rock Mech. Eng.* 2012, *31*, 1184–1191.

25. Wang, S.; Yang, J.; Yang, Y.; Zhang, F. Construction of large-span twin tunnels below a high-rise transmission tower: A case study. *Geotech. Geol. Eng.* 2014, *32*, 453–467.

26. PLAXIS 3D. *Foundation PLAXIS 3D Foundation User Manual, Version 2.0*; Brinkgreve, R.B., Swolfs, W.M., Eds.; PLAXIS Inc.: Delft, The Netherlands, 2008.

27. Liang, F.Y.; Chen, L.Z.; Shi, X.G. Numerical analysis of composite piled raft with cushion subjected to vertical load. *Comput. Geotech.* 2003, *30*, 443–453.

28. Hansbo, S. *Foundation Engineering*; Elsevier Science B.V.: Amsterdam, The Netherlands, 1994; pp. 89–91.

29. Jamiolkowski, M.; Ladd, C.C.; Germaine, J.T.; Lancellotta, R. New developments in field and laboratory testing of soils, theme lecture. In Proceedings of the 11th International Conference on Soil Mechanics and Found Engineering, San Francisco, CA, USA, 1985; Volume 1, pp. 57–153.

30. Mersi, G. Reevaluation of $s_{u(mob)}=0.22\sigma'_p$ using laboratory shear tests. *Can. Geotech. J.* 1989, *26*, 162–164.

31. Skempton, A.W. Discussion on the planning and design of new Hong Kong airport. *Proc. Inst. Civ. Eng.* 1957, *7*, 305–307.

32. Wroth, C.P. Interpretation of in-situ soil tests. *Geotechnique* 1984, *34*, 449–489.

33. Coduto, D.P. *Foundation Design—Principles and Practices*; Prentice Hall: Englewood Cliffs, NJ, USA, 1994.

34. Terzaghi, K.; Peck, R.B. *Soil Mechanics in Engineering Practice*; John Wiley & Sons: New York, NY, USA, 1967.

CHAPTER 2

3D-Analysis of Soil-Foundation-Structure Interaction in Layered Soil

Mohd Ahmed[1*], Mahmoud H. Mohamed[2], Javed Mallick[3], Mohd Abul Hasan[4]

Civil Engineering Department, Faculty of Engineering, King Khalid University, Abha, KSA

INTRODUCTION

The analysis of Piled-Raft Foundation is very challenging because the load in the piled-raft structures is transferred to the soil not only by the interaction between the soil and the piles but also by the interaction between foundation structure and superstructure. In this interaction, deformations in the soils are the key factor which will affect forces and deformation in foundation and superstructure. The soils below the ground level are heterogeneous and often found as layered system, i.e. layer wise varying properties below the ground. The combined piled-raft foundation penetrates deep into the foundation soil increasing its significant depth below the ground and affects the response of structure and soil. The method of analysis of foundation and structure also affects the response of structure and soil. The complex foundation system requires a reliable advance computational method that can simulate the 3D-non-linear soil behavior and structure-foundation system interaction. Considerable attention has been paid to analyze, design and construction of combined piled-raft foundation (CPRF) system. The survey of various analytical methods and numerical methods used to model the behavior of geomechanics has been presented by [1]. The various aspects contributed in reference to piled-raft foundation design have been compiled by Hemsley [2]. Ahmed et al. [3] has pointed out the recent advances in the piled-raft foundation system. Lin and Feng [4] have presented piled-raft analysis output for settlement, bending moment both in pile and raft, and effects of raft flexibility for vertical uniform loading in the subsoil. For the case of piled raft placed over soft clay layer, the contact pressure is merely 4% - 6%, whereas it is 15% - 25% if the piled raft resting on sand layer at ground surface. Rabiei [5] has carried out the parametric study on piled-raft foundation design. The parameter studied were pile length and spacing, number of piles, raft thickness, pile-soil and raft-soil stiffness ratio and pile-raft interaction. They concluded that by ignoring the interactions involved in the piled raft system, may lead to serious underestimates of settlement and also lead to inaccurate estimates of raft bending moments and pile loads. Singh and Singh [6] demonstrated that ignoring the interactions between the piled raft foundations elements may lead to a very serious over-estimate of the stiffness of the foundation. The case studies on optimized piled-raft foundation performance comprising of connected and non-connected piles using simple 2D analysis are presented by Eslami et al. [7]. A simplified procedure applicable has been presented by Kapackci and Ozkan [8] for estimation of piled-raft settlement. Nguyen et al. [9] has proposed a simplified design approach of piled-raft foundations under vertical load considering interaction effects. They compared the results of method with experimental and other numerical results and found good agreement between the results. The optimization study of piled-raft foundation systems has been carried out by Horikoshi and Randolph [10]. It is experimentally demonstrated that model rafts, founded on structurally disconnected pile reinforced sand, will have reduced settlement and bending moments [11]. Field measurements of the load observed for the raft and the piles of piled-raft foundation on stiff clays at working conditions are reported by Cooke [12]. They suggest that the ratio of load in the most heavily loaded piles in the perimeter of the group to that in the least heavily loaded pile near the centre could be about 2.5. A displacement based design procedure is proposed by Prakoso and Kulhawy [13] for piled-raft foundation based on the results of simplified linear elastic and nonlinear plane strain piled-raft finite element models. The effect of raft and pile group compression capacity was evaluated on the raft settlements, raft bending moments, and pile-raft load transfer ratio.

Mahmood and Ahmed [14] have carried out the dynamic analysis of framed including the soil-structure interaction effects and concluded that the soil-struc- ture interaction problem can have beneficial effects on the structural behavior when non-linear soil models and interface conditions are considered. Shayea and Zeedan [15] have presented a new approach for the design of raft foundation using 3-D modelling of each part of the whole structure (superstructure, raft and the soil) and considering the soil structure interaction. They developed charts to show the relationship between thickness of raft and number of design parameters including soil type.

From the literature survey it is clear that the interaction of the superstructure in the soil-foundation analysis has not been taken into consideration in most of the research work and load from the super structure in considered acting directly on the raft as a uniform or concentrated load. The effect of construction phase and mode of superstructure loading on the response of structure and foundation has not been given due attention. In this paper, complete soil-structure interaction of combined Piled-Raft Foundation with the foundation soil and superstructure of the building is evaluated through 3D-nonlinear Finite Element Analyses using PLAXIS3D foundation code [16] . The different parameters affecting the soil-foundation-structure response, such as building aspect ratios, mode of load application to foundation soil, soil failure criteria of soil field and proportional thickness ratio of stiff soil in two-layer soil stratum is studied. The displacements and load demands imposed on the high-rise building structures having piled-raft foundation are computed. The most of the previous studies on soil-piled-raft foundation analysis are based on direct loading of superstructure on raft and without considering interaction of superstructure and foundation. The foundation soil in piled-raft foundation-soil models without including super structure will be stiffer than models with the one-phase super structure loading or sequential super structure loading. The foundation structure and soil field response is significantly affected by different building structure shape and soil failure models. The soil field response in layered soil is also affected by presence of lesser stiff layer below the raft. It is also observed effect on deflection and forces of superstructure components due to inclusion of loading phases in piled-raft foundation interaction analysis.

STATEMENT OF THE PROBLEM

A 15-storey square/rectangular building having piled raft foundation in the two layered soil system is selected for the complete structure-foundation interaction analysis. The square building (aspect ratio = 1) has 4 bays in X- and Y-direction and the rectangular building (aspect ratio = 1.75) has 3 bays in X-direction and 6 bays in Y-direction as shown in Figure 1. Buildings have nearly the same plan area. The ground and typical floors are 6.0 m and 3.0 m high respectively. The structural system of all floors is a flat concrete slab type of 200 mm thickness subjected to a total uniform load of 15 kN/m². Dimensions of columns are listed inTable 1. The concrete raft is assumed to be at a depth 2.0 (m) beneath the ground surface and has 1.5 (m) thicknesses. The plan dimension of square raft is 25 m × 25 m with overhang of 2.5 m while the plan dimension of rectangular raft is 19.2 m × 33.6 m with overhang of 2.4 m. The estimated total vertical load on square and rectangular rafts is 138.8 MN. A total of 25 circular concrete piles of 0.75 m diameter are located under the raft for each building structure. Modulus of elasticity of concrete is assumed as 3.4 × 107 kN/m² while concrete Poisson's ratio and density is considered in structural models as 0.2 and 25 kN/m³ respectively. The typical floor plans and foundation plans having raft with pile location of square and rectangular shaped building are shown in Figure 1.

The slenderness ratio (L/D) of piles is taken as 26.7 and end tip of the piles are considered resting on the bottom surface of top soil layer having hardening soil model for different aspect of building, mode of application of structure loading on foundation and for different failure model of soils. Modulus of elasticity of pile material is taken as 2.35 × 107 kN/m² while its density is considered as 25 kN/m³. The soil profile is of two layer systems with upper layer of loose sand and lower layer of dense sand (stiff soil). The different thickness of stiff soil is considered in the model to study the effect of stiff soil on the interaction analysis. Three thickness proportion of stiff soil namely 25%, 50% and 75% of total thickness are taken. The water level is assumed at the ground surface.

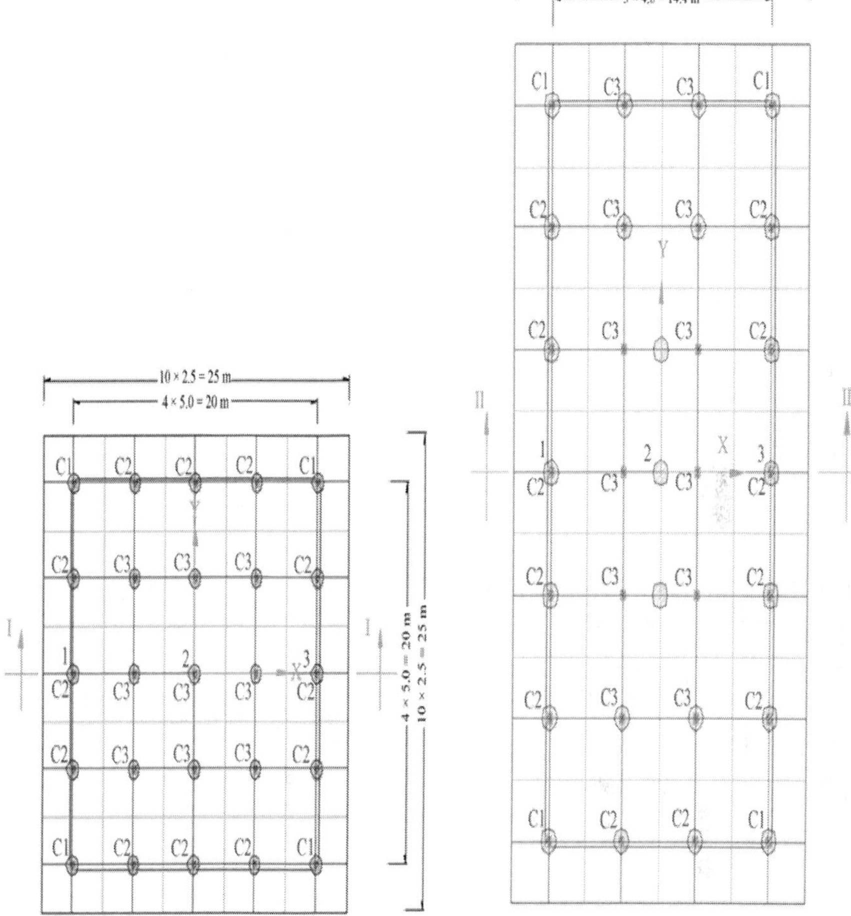

Figure 1. Typical Floor and Raft-Pile Plans of Buildings.

Table 1.Dimension Of Columns In Buildings.

Building Shape	Column Dimensions (m × m)		
	C1	C2	C3
Square	0.65 × 0.65	0.75 × 0.75	0.9 × 0.9
Rectangular	0.65 × 0.65	0.75 × 0.75	0.9 × 0.9

SOIL MODELS

Soil is a complex material that behaves differently in primary loading, unloading and reloading. It exhibits non- linear behaviour well below failure condition with stress dependant stiffness [17]. The elastic-perfectly plastic models based on soil failure criteria namely, Mohr-coulomb (MC) and Mohr-coulomb incremental stiffness (MCI) with depth, are taken because of the most common used models. In the study, second order Hardening Soil (HS) model that considers shear hardening and compression hardening [17] and suitable to cohesion less soil is also used. The assigned soil parameters for the two layers soil field are given in Table 2.

Finite Element Modelling Methodology

The finite element method based on software PLAXIS 3D is used for three dimensional modelling of 15-storey building structure having piled-raft foundation in layered soil field. The columns and piles are modelled as frame elements with linear elastic properties. The interaction effect of pile and soil at the pile shaft is considered by means of Elasto-Plastic line-to-volume and point-to-volume interfaces [19] as an embedded pile model. The embedded pile model consisting of beam elements with non-linear skin and tip interfaces. There is no need for mesh refinement around piles as 3D mesh is not distorted by introducing embedded pile model [19]. The floor slab and raft is discretized using 6-node triangular plate elements with linear elastic properties. The soil field with two layers of non-cohesive soils namely loose and dense sand is modelled as 15-node wedge triangular continuum elements. The total number of elements in the discretized mesh for square building structure including soil field are 35,785 while for rectangular building structure including soil field, number of elements are 24,785. As the mesh discretization has no significant effect on piled-raft analysis result [20], mesh sensitivity has not been examined and mesh generated automatically by PLAXIS code is used for the analysis. The 3D finite elements structural models of square (aspect ratio = 1) and rectangular (aspect ratio = 1.75) shaped buildings with piled-raft foundation are shown in Figure 2. The cross-section of 3D finite elements soil field models are shown in Figure 3.

Table 2. Soil Parameters for Different Soil Models.

Parameters	Soil model	Soil layer	
		Loose sand	Dense sand
Unsaturated weight (γ_{unsat}), kN/m^3	All models	17	19
Saturated weight (γ_{sat})	All models	20	21
Stiffness (E_{50}^{ref}), kN/m^2	HS [9]	20,000	60,000
Stiffness (E_{oed}^{ref}), kN/m^2	HS [9]	20,000	60,000
Stiffness (E_{ur}^{ref}), kN/m^2	All models	100,000	180,000
Rate of increase of E with depth (ΔE) kN/m^2	MCI [18]	4720	31,470
Power (m)	HS	0.65	0.55
Poisson's ratio (ν)	All models	0.2	0.2
Dilatancy (ψ), degree	All models	2	8
Friction angle (φ), degree	All models	32	38
Cohesion (c_{ref}), kN/m^2	HS	0.1	0.1

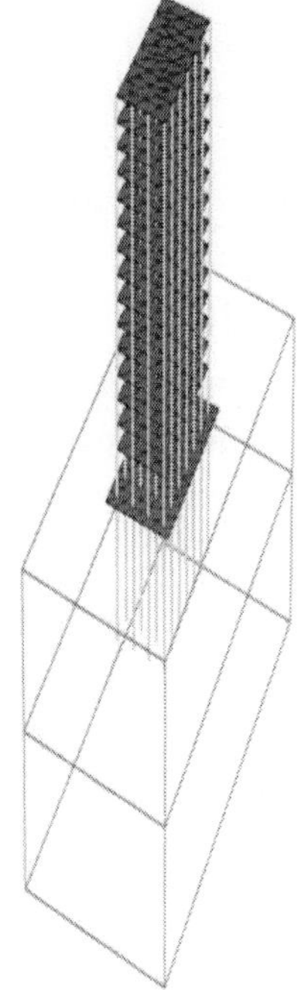

Figure 2. 3d Model of Buildings with Piled-Raft Foundation.

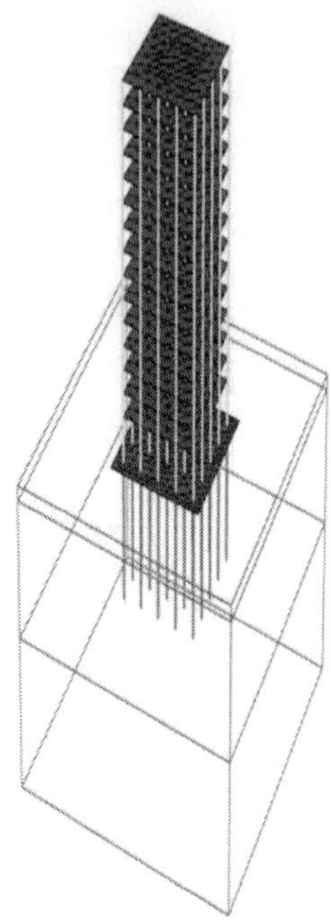

Figure 3. 3d Finite Elements Foundation Soil Models Of Square And Rectangular Shaped Buildings.

RESULTS AND DISCUSSION

Foundation-Structure Interaction Effect on Soil Field

The results of interaction of building foundation-structure with different aspect ratio of building and different soil models on the soil are given in Table 3. The results are presented at three locations below the ground level, i.e. 2 m (below raft level), 10 m and 20 m below the ground. It is evident from the results that there is noteworthy effect of interaction of building aspect ratio and soil failure models on the soil response. The predicted amount of soil settlement is different with different aspect ratio of building. The soil settlement decreases with

Table 3. Maximum Soil Settlement with Different Soil Models and Building Aspect Ratio

Soil Models	Location		Maximum Soil Settlement (cm)
Hardening Model (Aspect Ratio = 1.0 & 1.75)	Below the raft, 2 m below GL	Aspect Ratio = 1.0	23.19
		Aspect Ratio = 1.75	22.0
	10 m below GL	Aspect Ratio = 1.0	11.74
		Aspect Ratio = 1.75	11.30
	Below the pile, 20 m below GL	Aspect Ratio = 1.0	3.22
		Aspect Ratio = 1.75	3.05
Mohr-Coulomb Incremental Model (Aspect Ratio = 1.0)	Below the raft, 2 m below GL		26.83
	10 m below GL		14.37
	Below the pile, 20 m below GL		4.16
Mohr-Coulomb Incremental Model (Aspect Ratio = 1.0)	Below the raft, 2 m below GL		12.97
	10 m below GL		4.43
	Below the pile, 20 m below GL		1.23

the increase of aspect ratio of building. This may due to the reason that the load is distributed on a larger area in one direction of the building. The use of different soil failure model for soil field has also predicted dissimilar soil settlement. The behavior of soil at various levels also varies under different failure models of soil. The soil settlement is predicted highest using Mohr-coulomb failure criteria (MC) and predicted least by Mohr-coulomb incremental stiffness model (MCI). The soil settlements of a square building (aspect ratio = 1) at raft level are 26.83 cm, 12.97 cm and 23.19 cm respectively in Mohr-coulomb (MC) model, Mohr-coulomb incremental stiffness (MCI) model and hardening soil (HS) model while, the settlements at pile end are 4.16 cm, 1.23 cm and 3.22 cm. For the rectangular structure (aspect ratio = 1.75), the soil settlement is concentrated in the shorter direction of the building structure. The contours of soil settlements along vertical cross-section of the soil field for building aspect ratios and soil failure criteria are depicted in Figure 4 and Figure 5.

Table 4 presents the soil settlements variation with proportional depth of loose sand layer of the building foundation soil. The contours of soil settlements along vertical and horizontal cross-section of the soil field are shown in Figure 6. It is evident from the results that the maximum soil settlements go on increasing with the increase of depth of loose sand layer below the ground level. The soil settlements are also increases with the increase of loose sand layer at same level below the ground level. The maximum soil settlements with different proportional depth of loose sand layer are 18.46 cm, 23.19 cm and 23.95 cm respectively at loose sand depth 25%, 50% and 75% of total depth, while with similar proportional depth of loose sand layer, the settlements at bottom pile end are 2.76 cm, 3.22 cm and 6.0 cm. For different proportional depth of loose sand layer, settlement dissipates along the raft sides towards the outer edges of soil field. The change in soil settlement is observed with lesser stiff layer thickness up to the pile length and more thickness of lesser stiff layer will not affect the soil behavior in piled-raft foundation.

Table 5 shows the predicted soil settlements with the mode of super structure loading to the foundation soil. The contours of soil settlements along the vertical and horizontal cross-section of the soil field are shown in Figure 7. It is evident from the results that there is clear interaction of building structure with the foundation soil. The analysis of piled-raft foundation-soil models with super structure will indicate foundation soil to be more rigid than model without the super structure. The maximum soil settlements will decreases when loading is applied through the construction phasing of the building structure or sequential loading. The maximum soil settlements will increase when loading is applied through the vertical elements tributary area method of the structure. The maximum soil settlements are 23.19 cm 23.45 cm and 22.74 cm respectively at loading through the super structure, loading directly to footing and through sequential loading while on similar conditions of loading, the soil settlements at bottom pile end are 3.22 cm, 3.25 cm, and 3.16 cm.

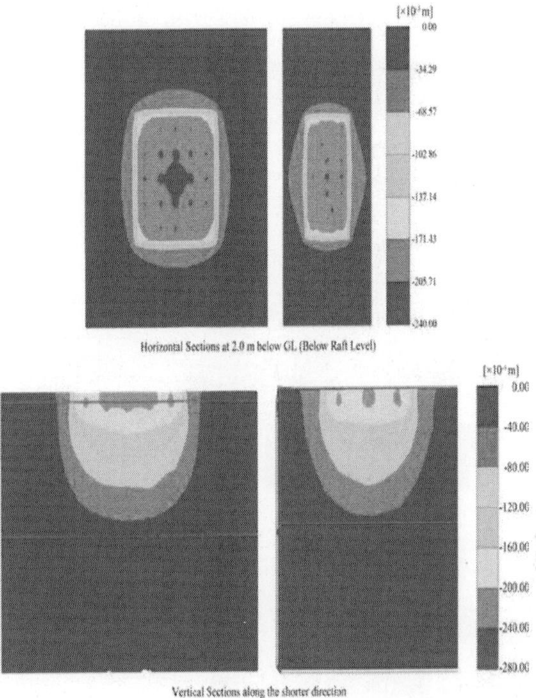

Figure 4. Contours of Soil Settlements On Horizontal And Vertical Section Of Square (Aspect Ratio = 1.0) And Rectangular Buildings (Aspect Ratio = 1.75),

Table 4. Maximum Soil Settlement of Square Structure Building with Loose Sand Layer (Hs) Proportional Depth.

Soil Models	Location	Maximum Soil Settlement (cm)
Top Loose Sand—−25% of total depth	Below the raft, 2 m below GL	18.46
	10 m below GL	5.49
	Below the pile, 20 m below GL	2.76
Top Loose Sand—−50% of total depth	Below the raft, 2 m below GL	23.19
	10 m below GL	11.74
	Below the pile, 20 m below GL	3.22
Top Loose Sand—−75% of total depth	Below the raft, 2 m below GL	23.95
	10 m below GL	13.89
	Below the pile, 20 m below GL	6.0

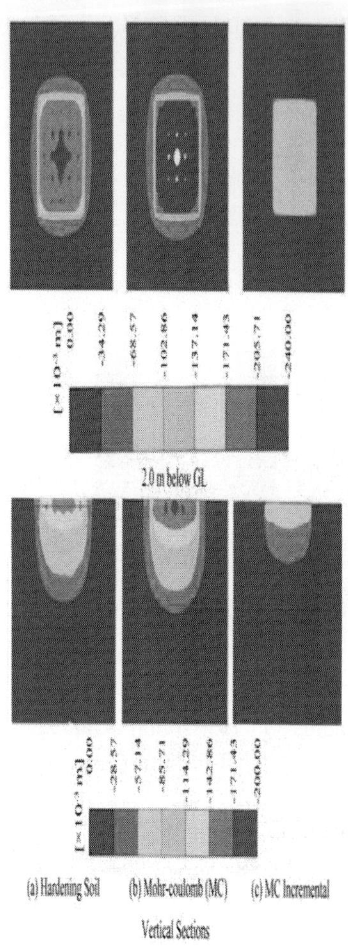

Figure 5. Contours Of Soil Settlements On Horizontal And Vertical Section Of Square Buildings With Soil Models.

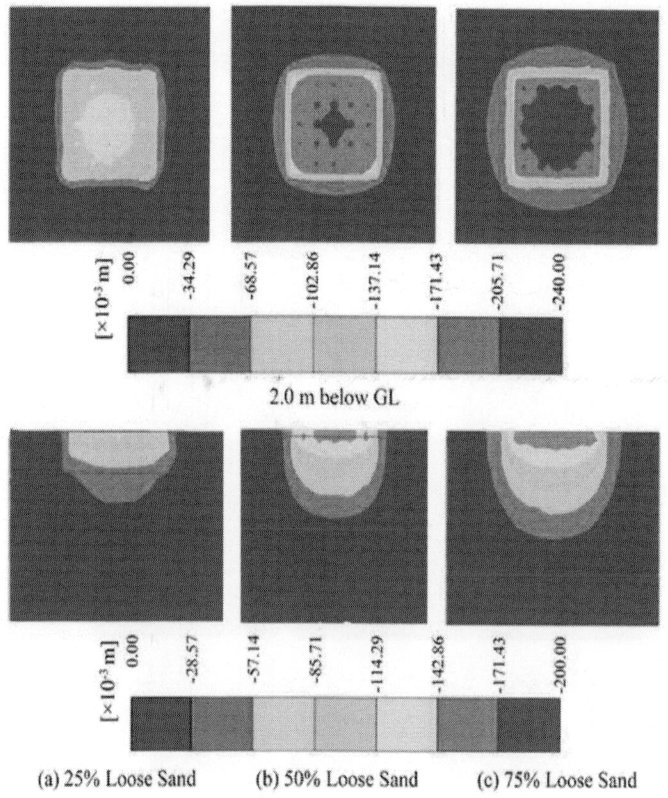

Figure 6. Contours of Soil Settlements on Horizontal And Vertical Section Of Square Buildings With Proportional Depth Of Loose Sand (HS) Layer.

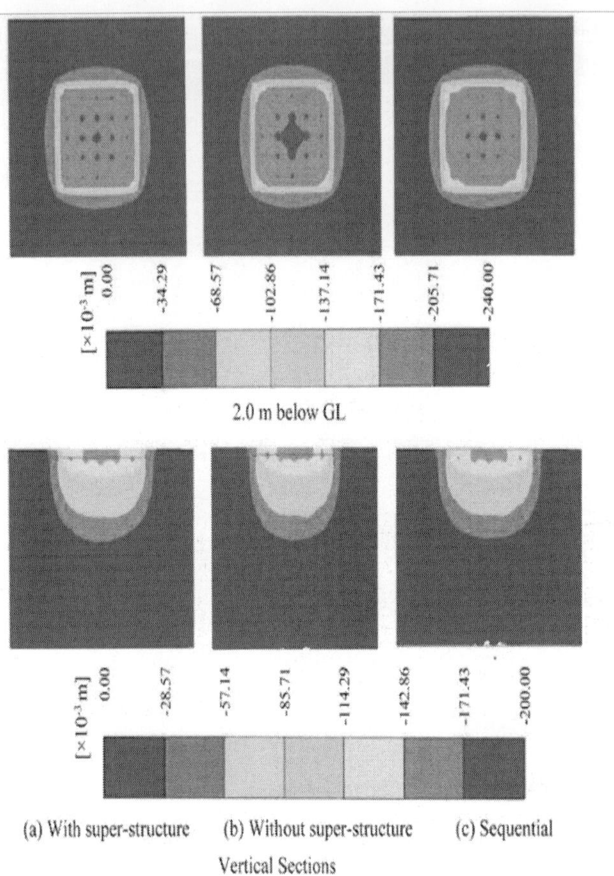

Figure 7. Contours Of Soil Settlements On Horizontal And Vertical Section Of Squarebuildings With Mode Of Super Structure Loading.

Table 5. Maximum Soil Settlement of Square Structure Building With Mode Of Super Structure Loading.

Soil Models	Location	Maximum Soil Settlement (cm)
Loading through super structure	Below the raft, 2 m below GL	23.19
	10 m below GL	11.74
	Below the pile, 20 m below GL	3.22
Loading without super structure	Below the raft, 2 m below GL	23.45
	10 m below GL	11.6
	Below the pile, 20 m below GL	3.25
Sequential loading through super structure	Below the raft, 2 m below GL	22.74
	10 m below GL	11.47
	Below the pile, 20 m below GL	3.16

Foundation-Structure Interaction Effect on Piled-Raft Footing

Table 6 shows the analysis results of interaction of building foundation-structure on the component of piled-raft footing. The table shows the computed settlements of raft and total static load transferred from the upper structure to the raft of building with different aspect ratios and soil failure models. The forces developed in the raft are also presented in this table. The results of analysis depicts that the amount of raft maximum and differential settlement vary with different aspect ratio of buildings and failure models of soil. The differential settlement and raft forces decrease with the increase of aspect ratio of the building structure. The predicted differential settlement of the raft is highest using Mohr-coulomb failure criteria of soil field and it is least in Mohr-coulomb incremental stiffness model. The least value of maximum positive and negative moment in the raft is computed

Table 6. Differential Settlement, Moments And Forces In The Raft With Different Soil Models And Building Structure Aspect Ratio.

Building Aspect Ratio	Settlements/Max. Moments/Max. Shear Force/Vertical Load	Soil Models		
		Hardening Model	Mohr Coulomb (MC) Model	MC Incremental Model
Aspect Ratio = 1.0 (Square)	Differential Settlement (cm)	7.3	7.8	6.37
	Positive Moment (kN·m/m)	4850	5092	4455
	Negative Moment (kN·m/m)	130	121	120
	Shear Force (kN)	4852	4986	4916
	Total Vertical Load (kN)	138,801	138,801	137,014
Aspect Ratio = 1.75 (Rectangular)	Differential Settlement (cm)	6		
	Positive Moment (kN·m/m)	2562		
	Negative Moment (kN·m/m)	213		
	Shear Force (kN)	1401		
	Total Vertical Load (kN)	138,812		

with Mohr-coulomb incremental stiffness (MCI) failure criteria. The computed value of raft maximum positive moments are (4850, 5092, 4455 kN·m/m), raft maximum negative moments are (130, 121, 120 kN·m/m), and raft maximum shear force are (4852, 4986, 4916 kN) using hardening soil (HS) model, Mohr-coulomb (MC) model and Mohr-coulomb incremental stiffness (MCI) model respectively. The maximum positive moment, maximum negative moment and maximum shear force in the raft obtained from the interaction analysis are 2563 kN·m/m, 213 kN·m/m and 1401 kN respectively for building aspect ratio of 1.75.

Table 7 and Table 8 show the differential settlement and forces developed in the raft with proportional depth of loose sand and mode of loading to the foundation soil. The results of analysis concluded that the amount of raft maximum and differential settlement varies with different proportional depth of loose sand and mode of loading to the foundation soil. The differential settlement of the raft decreases with the increase of proportional depth of loose sand. The raft forces decreases with the increase of aspect ratio of the building structure. The predicted differential settlement of the raft is highest using Mohr-coulomb failure criteria of soil field and it is least in Mohr-coulomb incremental stiffness model. The lowest value of maximum negative moment in the raft is computed with lower thickness of loose sand layer while the lowest value of maximum shear in the raft is obtained with highest thickness of loose sand layer. The computed value of raft maximum positive moments are (5129, 4850, 4809 kN·m/m), raft maximum negative moments are (117, 130, 129 kN·m/m), and raft maximum shear force are (2390, 4852, 2295 kN) using 25%, 50% and 75% proportional thickness of loose sand layer respectively.

The analysis results of piled-raft foundation model, developed without the superstructure and loading directly to structure based on tributary area of columns and loading through super-structure with or without phasing of construction, is shown in the Table 8. The results of piled-raft foundation model with and without the superstructure indicate a clear interaction between the foundation-soil and the super structure. The bending moments and shear force due to loading and differential settlements in the raft are lesser in case of piled-raft foundation-soil model without the building super structure. The developed raft bending moments is maximum when is loading is transferred to footing is sequential manner. There is no noticeable interaction effect on differential settlements of the raft with different mode of application of loading to the foundation. The maximum and differential settlements of soil field due to piled-raft foundation with different mode of loading to foundation are (23.19 cm, 7.3 cm) and (23.45 cm, 6.76 cm), (22.74 cm, 7.32 cm) respectively due to loading through building super structure, loading directly to footing and due to sequential loading of the super structure. The maximum positive moments and negative moments, and maximum shear force in the raft are (4850 kN·m/m, 130 kN·m/m, 4852 kN), (3916 kN·m/m, 170 kN·m/m, 3626 kN) and (4850 kN·m/m, 131 kN·m/m, 4992 kN) respectively due to loading through building super structure, loading directly to piled-raft footing and due to sequential loading of

Table 7: Settlement, Moments And Forces In The Raft With Proportional Depth Of Loose Sand (HS) Layer

Settlements/Max. Moments/Max. Shear Force/Vertical Load	Proportional Depth of Loose Sand (HS) Layer		
	25% Loose Sand	50% Loose Sand	75% Loose Sand
Differential Settlement (cm)	9.26	7.3	7.26
Positive Moment (kN·m/m)	5129	4850	4809
Negative Moment (kN·m/m)	117	130	129
Shear Force (kN)	2390	4852	2295
Total Vertical Load (kN)	138,759	138,801	138,767

Table 8: Settlement, Moments And Forces In The Raft With Mode Of Super Structure Loading.

Settlements/Max. Moments/Max. Shear Force/Vertical Load	Mode of Super Structure Loading		
	Single Phase Super-Structure Loading	Without Super-Structure-Direct Loading	Sequential Loading
Maximum Settlement (cm)	23.19	23.45	22.74
Differential Settlement (cm)	7.3	6.76	7.32
Positive Moment (kN·m/m)	4850	3916	4856
Negative Moment (kN·m/m)	130	170	131
Shear Force (kN)	4852	3626	4992
Total Vertical Load (kN)	138,801	138,808	138,804

Foundation-Structure Interaction Effect on Super-Structure

Table 9 and Table 10 show the results of interaction of building foundation-structure on the component of super-structure (i.e. columns and slabs) due to phasing of construction. The side sways (deflections) and axial loads in the columns are given in Table 9. The axial loads in columns of building are affected by the different mode of loading applied to the foundation soil. The maximum axial loading in the column increases while minimum axial loading decreases due to application of loading in sequential way. The maximum side sway or deflection of the column is increased when applied loading considers the construction phasing. The maximum top side sway in x and y directions for sequential loading to the building structure are 95 mm and 207 mm while deflections for one phase loading to the foundation are 12.2 mm and 43.3. The interaction of the super structure with the raft-foundation-soil will affect the maximum and minimum axial load of the columns. The maximum and minimum axial loads of the columns, obtained by incorporating the super-structure mode of loading to piled- raft foundation-soil model, are (9938 kN, 1768 kN) and (10,244 kN, 1740 kN) for one phase loading and for sequential loading to the building structure respectively while based on the column tributary area, the maximum and minimum axial load of the columns are 7556 kN and 2268 kN.

Table 10 depicts the maximum deflections and moments in slab at first, eighth floor and roof of the super structure with sequential loading to the foundation. The floor slab deflections and slab forces are different with different mode of loading of the building super structure. There is clear interaction effect on different floor of the building with phase loading of construction. The maximum deflection of the first floor slab is observed highest due to loading of super structure applied in sequential manner but observed decreasing on upper floors of building structure while deflection observed is more or less same if super structure loading is applied in single phase. The maximum deflection produced in the floor slabs of roof, eighth and first floor of the building super structure with different mode of loading to the foundation are (38.6 cm, 37.5 cm, 34.6 cm) and (16.1 cm, 24.0 cm, 38.8 cm) respectively due to loading in one phase through building super structure and due to sequential loading of the super structure. The moments developed in the roof slab are observed more after application of sequential loading of super structure than the other floor slabs and go on increasing on others floors of building structure. The maximum positive moments and negative moments in the roof slab, eighth and first floor of the

Table 9. Columns Deflection and Axial Loads With Mode Of Super Structure Loading.

Top Deflection (mm)/Axial Load (kN)	Mode of Super Structure Loading		
	Super-Structure Loading	Sequential Loading	Without Super-Structure Model
Top Deflection (x-dir.)	23.19	23.45	22.74
Top Deflection (y-dir.)	7.3	6.76	7.32
Axial Load (max.)	4850	3916	4856
Axial Load (min.)	130	170	131

Table 10. Maximum Deflection and Moments in the Floors of Square Building Structure with Different Mode Of Super Structure Loading.

Storey Level	Max. Slab Deflection/Max. Moments	Mode of Super Structure Loading	
		Single Phase Super-Structure Loading	Sequential Loading
First Floor	Deflection (mm)	34.6	38.8
	Positive Moment (kN·m/m)	64	64
	Negative Moment (kN·m/m)	97	97
Eighth Floor	Deflection (mm)	37.5	24.0
	Positive Moment (kN·m/m)	65	63
	Negative Moment (kN·m/m)	95	100
Roof	Deflection (mm)	38.6	16.1
	Positive Moment (kN·m/m)	66	61
	Negative Moment (kN·m/m)	94	106

building are respectively (66 kN·m/m, 94 kN·m/m), (65 kN·m/m, 95 kN·m/m) and (64 kN·m/m, 97 kN·m/m) due to loading through building super structure and while these forces respectively are (61 kN·m/m, 106 kN·m/m), (63 kN·m/m, 100 kN·m/m) and (64 kN·m/m, 97 kN·m/m) due to sequential loading of the super structure.

CONCLUSIONS

The analysis of combined piled-raft foundation of multi-storey building is very challenging because of complexities involved in the interaction between the components of building structure and soil field. The analysis of the tall building structure with complex foundation system in non-uniform (layered soil) soil field should include the interaction of structure-foundation-soil. In this study, the finite element 3D interaction analysis of building structure having piled-raft foundation in two layered non-cohesive soil field is carried out using PLAXIS 3D foundation code. The complete interaction among the soil field depth, soil layer type with foundation and foundation with super-structure with different aspect ratio and loading mode has been evaluated. The available literatures on soil-piled-raft foundation analysis are based on direct loading of superstructure on raft and without considering interaction of superstructure and foundation. The foundation soil in piled-raft foundation-soil models without including super structure will be stiffer than models with the one-phase super structure loading or sequential super structure loading.

The foundation structure and soil field response is significantly affected by different building structure shape and soil failure models. The foundation soil settlement and raft differential settlement is highest using Mohr-coulomb (MC) failure criteria of soil field among the HS, MC and MCI failure criteria. The soil field response in layered soil is also affected by presence of lesser stiff layer below the raft. The soil behavior in piled-raft foundation is not much affected by lesser stiff layer having thickness more than the pile length. A clear foundation- structure interaction effect is observed on the building superstructure components behavior with application of construction loading sequentially. The wide variability of deflection and moments of the floor slab is also observed due to loading of super structure applied in sequential manner which is not observed when super structure loading is applied as a single phase. The deflection and moments of the first floor slab is observed highest due to loading of super structure applied in sequential manner but observed decreasing on upper floors of building structure.

REFERENCES

1. Ahmed, M., Mahmoud, H. and Mallick, J. (2013) Advances in Piled-Raft Foundation System. Recent Trends in Civil Engineering and Technology, 3, 1-8.

2. Al-Shayea, N. and Zeedan, H. (2012) A New Approach for Estimating Thickness of Mat Foundations under Certain Con- ditions. The Arabian Journal for Science and Engineering, 37, 277-290. http://dx.doi.org/10.1007/s13369-012-0178-5

3. Bobet, A. (2010) Numerical Methods in Geo-Mechanics. The Arabian Journal for Science and Engineering, 35, 27- 48.

4. Cao, X.D., Wong, M.F. and Chang, M.F. (2004) Behavior of Model Rafts Resting on Pile-Reinforced Sand. Journal Geotechnical Engineering, 130, 129-138.

5. Cooke, R.W. (1986) Piled Raft Foundations on Stiff Clays: A Contribution to Design Philosophy. Geotechnique, 36, 169-203. http://dx.doi.org/10.1680/geot.1986.36.2.169

6. Engin, H.K. and Brinkgreve, R.B.J. (2009) Investigation of Pile Behavior Using Embedded Piles. Proceedings of the 17th International Conference on Soil Mechanics and Geotechnical Engineering. In: Hamza, M., et al., Eds., Alexandria, Millpress, Amsterdam.

7. Eslami, A., Veiskarami, M. and Eslami, M.M. (2012) Study on Optimized Piled-Raft Foundations (PRF) Performance with Connected and Non-Connected Piles-Three Case Histories. International Journal of Civil Engineering, 10, 100- 110.

8. Hemsley, J.A. (2000) Design Applications of Raft Foundations. Thomas Telford Ltd., London.

9. Horikoshi, K. and Randolph, M.F. (1998) A Contribution to Optimum Design of Piled Rafts. Geotechnique, 48, 301- 317. http://dx.doi.org/10.1680/geot.1998.48.3.301

10. Kapackci, V. and Ozkan, M.Y. (2012) A Simplified Approach Applicable to the Settlement Estimation of Piled-Raft. ACTA Geo-Technica Slovenica, 1, 77-84.

11. Lebeau, J.S. (2008) FE-Analysis of Piled and Piled Raft Foundations. Ph.D., Institute for Soil Mechanics and Foundation Engineering, Graz University of Technology, Graz.

12. Lin, D.G. and Feng, Z.Y. (2006) A Numerical Study of Piled Raft Foundations. Journal of the Chinese Institute of Engineers, 29, 1091-1097.http://dx.doi.org/10.1080/02533839.2006.9671208

13. Mahmood, M.N. and Ahmed, S.Y. (2007) Nonlinear Dynamic Analysis of Framed Structures including Soil-Structure Interaction Effects. The Arabian Journal for Science and Engineering, 33, 45-64.

14. Nguyen, D.D.C., Jo, S.B. and Kim, D.S. (2013) Design Method of Piled-Raft Foundations under Vertical Load Considering Interaction Effects. Computers and Geotechnics, 47, 16-27. http://dx.doi.org/10.1016/j.compgeo.2012.06.007

15. PLAXIS Version 2012.02 (2012) Scientific Manual, Delft University of Technology & PLAXIS, The Netherlands, A. A. Balkema, PUBLISHERS. http://www.plaxis.nl/

16. Prakoso, W.A. and Kulhawy, F.H. (2001) Contribution to Piled Raft Foundation Design. Journal Geotechnical Engineering, 127, 17-24.

17. Rabiei, M. (2010) Piled Raft Design for High Rise Building. Electronic Journal of Geotechnical Engineering, 15, 495- 505.

18. Singh, N.T. and Singh, B. (2008) Interaction Analysis for Piled Rafts in Cohesive Soils. 12th International Conference of International Association for Computer Methods and Advances in Geo-Mechanics (IACMAG), Goa, 1-6 October 2008, 3289-3296.

19. Ti, K.S., Huat, B.B.K., Noorzaei, J., Jaafar, M.S. and Sew, G.S. (2009) A Review of Basic Soil Constitutive Models for Geotechnical Application. Electronic Journal Geotechnical Engineering, 14, 1-18.

20. US Department of Transportation (2006) Federal Highway Administration. Publication No.: FHWA NHI-06-088.

CITATION

Ahmed, M, Mohamed, M , Mallick, J. and Hasan, M. (2014) 3D-Analysis of Soil-Foundation-Structure Interaction in Layered Soil. *Open Journal of Civil Engineering*, **4**, 373-385. doi: 10.4236/ojce.2014.44032.

CHAPTER 3

Experimental Study of Dynamic Characteristics on Composite Foundation with Cfg Long Pile and Rammed Cement-Soil Short Pile

Jihui Ding[1], Yanliang Cao[1], Weiyu Wang[2], Tuo Zhao[2], Junhui Feng[3]

[1]College of Civil Engineering, Hebei University, Baoding, China

[2]Hebei Academy of Building Research, Shijiazhuang, China

[3]China Metallurgical Design and Research Institute Co., Ltd., Baoding, China

INTRODUCTION

The composite foundation is that the part soil body in the natural ground foundation is reinforced or replaced during the ground treatment, and load is born by reinforced body and soil body around the pile [1]. Design theory of single pile composite foundation is relatively mature, and has certain limitations and shortcomings. The pile stiffness is smaller and the pile body has certain bond strength in the flexible pile composite foundation. The most commonly flexible piles are mixing cement soil pile [2] [3], rammed cement-soil pile [4] [5] pile, and so on. The strength of the flexible piles is low and load cannot be effectively transmitted to the lower part of the pile. When the top of the soil-cement pile is crushed, the side friction of the pile length range did not develop out.

The pile in the rigid pile (i.e. CFG pile) composite foundation has higher strength, with large adjustment range of bearing capacity of composite foundation [6]. Usually the rigid pile has happened with piercing failure, and the pile body material strength has not fully developed out. To give full play to the advantages of various types of pile, the composite foundation to form by different typed piles combined together [7] [8], can maximize the advantages of various types of pile. With the rapid development of economy, strength, length of pile in composite foundation can be greatly improved, and greatly improve the bearing capacity of composite foundation. The original design theory of composite foundation cannot meet the requirements, and dynamic problems of composite foundation have become the focus of attention. Study on the seismic performance of the composite foundation is the main application of numerical analysis and simulation of the composite foundation of finite element software [9] -[11] , this method still remain at the theoretical level, have not been applied to the actual design. Wang Weiyu, Zhao Tuo, Ding Jihui etc. studied dynamic characteristic and its influence factors of cement soil pile and CFG pile composite foundation under the action of the blasting vibration [12] -[15] .

Optimization is made to CFG pile and rammed cement-soil pile, and the composite foundations of CFG pile, CFG long pile and CFG short pile, CFG long pile and rammed cement-soil pile are, the designed. The stress and dynamic characteristics of the three composite foundation are studied through field tests.

THE INTRODUCTION OF THE TEST SITE

The test site is located in Shijiazhuang Heibei province. In the 20 m depth, the soil layers mainly are yellow silt clay, fine sand, middle sand and silt clay. In the 20 m driving depth, the underwater is not seen. There is not the harmful geologic action in the site. The main parameters of soil layer as shown in Table 1.

MODEL TEST AND SCHEME OF THE SITE

Three kinds of composite foundation model are designed: CFGP, CFGLP-CFGSP, CFGCP-RCSLP composite foundation. The model design parameter of composite foundation is shown in Table 2. CFG pile adopts C20 commercial concrete. Blasting is used as the vibration resource. The diameter of blasting hole is 50 mm. The Explosives are buried in the hole and then backfill tamping. The vibration is picked by acceleration sensors. The arrangement of the piles and measuring elements are shown in Figures 1-4.

The upper load is supplied by pilling concrete block, and each load of the composite foundation is added by an electric pressure pump-hydraulic jack. The square steel is 2.0 × 2.0 m as the loading plate.

THE ANALYSIS OF EXPERIMENTS RESULT

Load-Settlement Curves

Combining the three kinds load test of composite foundation, the load-settlement curves as shown in Figure 4. From Figure 4, compared with natural foundation, the bearing capacity of composite foundation of CFGP, CFGLP-CFGSP and CFGLP-RCSSP increases obviously and the deformation of composite foundation de- creased compared to the Natural Foundation. The nonlinear degree of the p - s curves of combined pile decreases, and CFGLP-RCSSP is closed to linear relation. The bearing capacity of the four composite pile of the CFGP, CFGLP-CFGSP, and CFGLP-RCSSP in the site are 225 kPa, 179 kPa, and 197 kPa, separately increases 150%, 98.8% and 119% compared to the Natural Foundation.

Table 1. Mainly Parameter of Soil Layer.

No.	h/m	f_{ak}/kpa	E_s/Mpa	f_{sk}/kpa	f_{pk}/kpa
1)	0.5	130	6.17	60	
2)	1.5	140	10	55	800
3)	4.0	200	11.5	82	1500
4)		270	5.84	65	1000

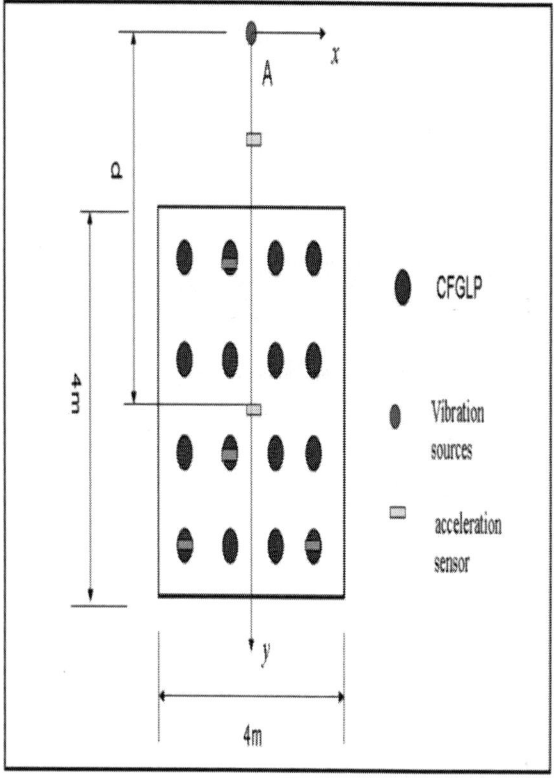

Figure 1. The Arrangement of The Piles And Measuring Elements Of The Model 1.

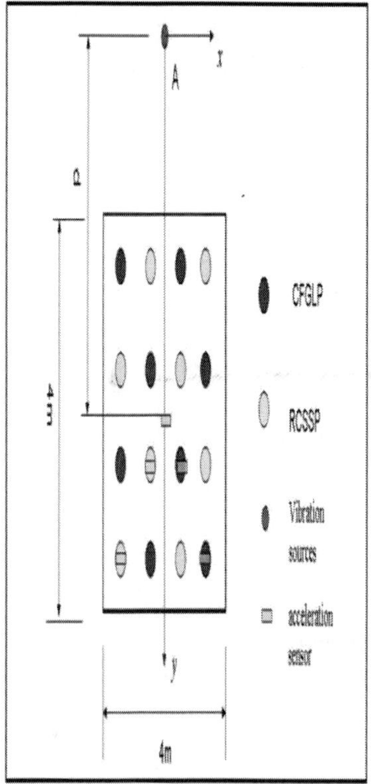

Figure 2. The Arrangement Of The Piles And Measuring Elements Of The Model 2.

Table 2. Model Design Parameter of Composite Foundation.

Model	Type	Pile length/m	Pile diameter/mm	Pile spacing/mm	Replacement rate
1	CFGP	6.0	350	100	0.09616
2	CFGLP	6.0	350	200	0.04808
	CFGSP	4.0	350	200	0.04808
3	CFGLP	6.0	350	200	0.04808
	RCSSP	4.0	350	200	0.04808

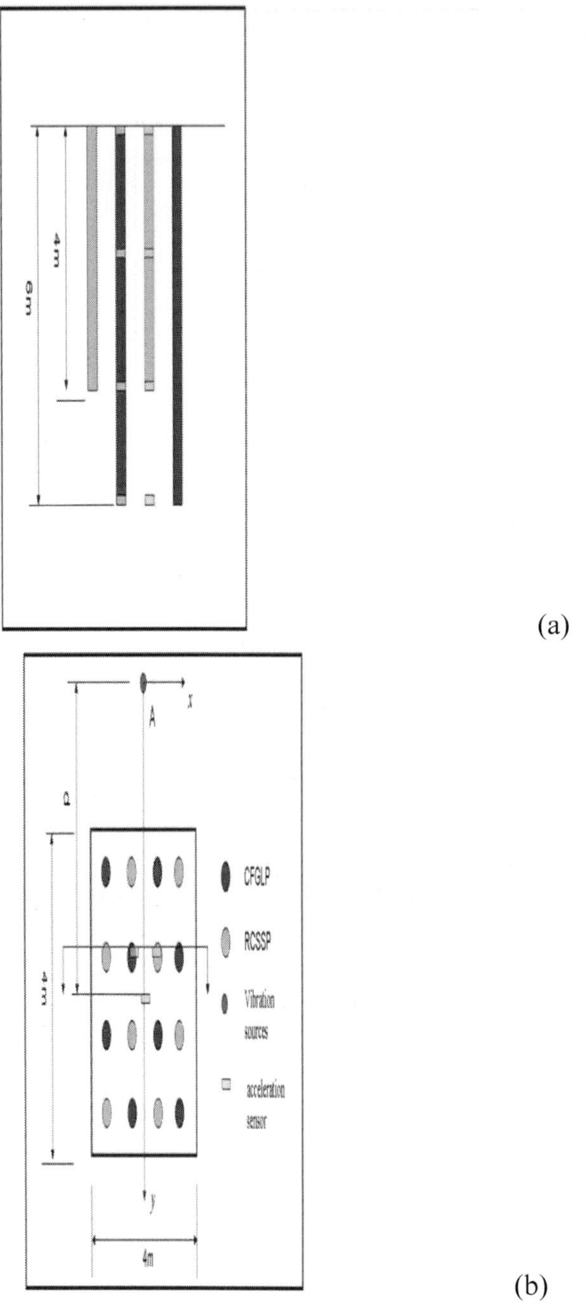

Figure 3. The Arrangement of The Piles And Measuring Elements Of The Model 3. (A) Plane Arrangement ;(B)1-1profilearrangement.

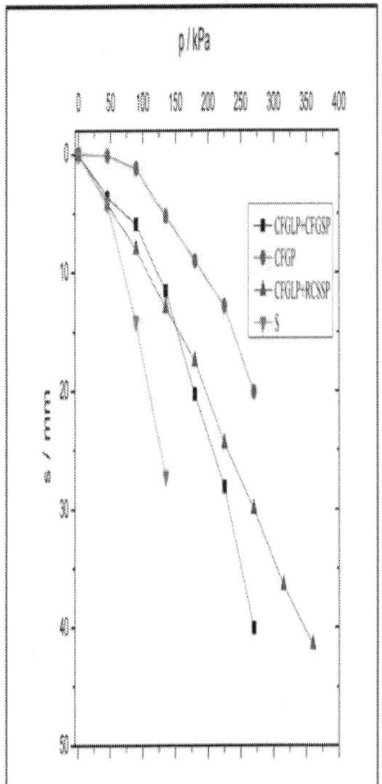

Figure 4. The P - S Curves Of The Composite Foundation.

The Main Frequency of Vibration

Field test shows that, under the same blast energy, vibration position, properties of foundation soil and form of composite foundation, have influence on the main frequency of composite foundation, but the rule is not obvious. Vibration main frequency depends on the properties of the soil and pile body between the vibration source and measuring point. When no load, the horizontal vibration main frequency of the natural foundation is at 8.2 - 9.03 Hz, and the vertical vibration main frequency is at 7.75 - 9.16 Hz; the horizontal vibration main frequency CFGP composite foundation is at 21.1 - 27.7 Hz, and the vertical vibration frequency is at 6.8 - 31.6 Hz; the horizontal vibration main frequency of the CFGLP-CFGSP composite foundation is at 17.4 - 29.6 Hz, and the vertical vibration main frequency is at 7.2 - 9.4 Hz; the horizontal vibration main frequency of the CFGLP-RCSSP composite foundation is at 7.6 - 38.4 Hz, and the vertical vibration main frequency is at 8.2 - 43.1 Hz. Horizontal vibration main frequency greater than the vertical vibration main frequency, and the vertical main vibration frequency close to the natural frequency of composite foundation. With the load increasing, the main vibration frequency of the CFGLP-RCSSP composite foundation decreases. When the load is 180 kPa, the horizontal vibration main frequency of the CFGLP-RCSSP composite foundation is at 7.4 - 22.6 Hz, and the vertical vibration main frequency is at 7.3 - 28.0 Hz.

Peak Acceleration Results

Figures 5-7 are the peak acceleration with the horizontal distance r from the measuring point to the vibration source without no pilling load on the CFGP composite foundation, when the depth of the vibration source is 6 m and the explosive quantity is 1.05 kg. From Figure 4, in addition to individual point the horizontal peak acceleration is greater than the vertical peak acceleration, and with the increase of r, the difference gradually decreases. From Figures 5-7, outside the scope of the composite foundation, the peak acceleration significantly decreased with the increase of the pilling load; in the surface of compound foundation, pilling load action makes the peak acceleration gently.

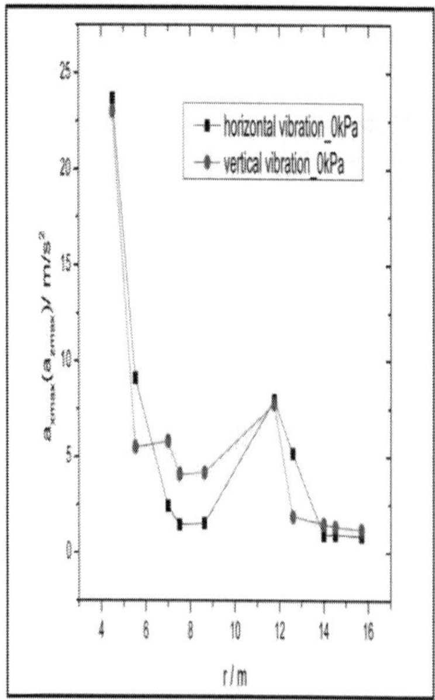

Figure 5. $A_{xmax}(A_{zmax})$-R Of CFGP Composite Foundation (No Load).

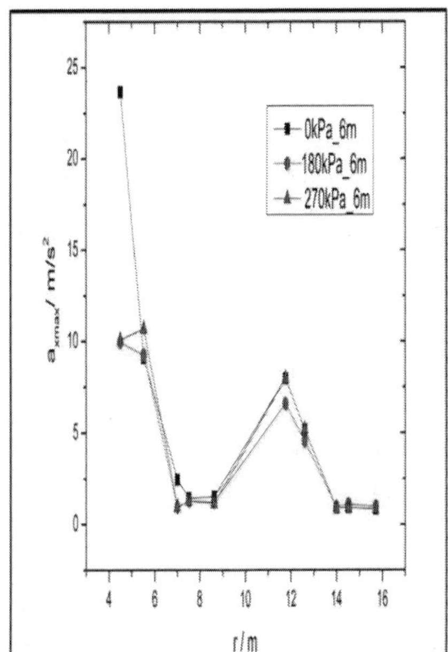

Figure 6. A_{xmax}-R Of CFGP Composite Foundation.

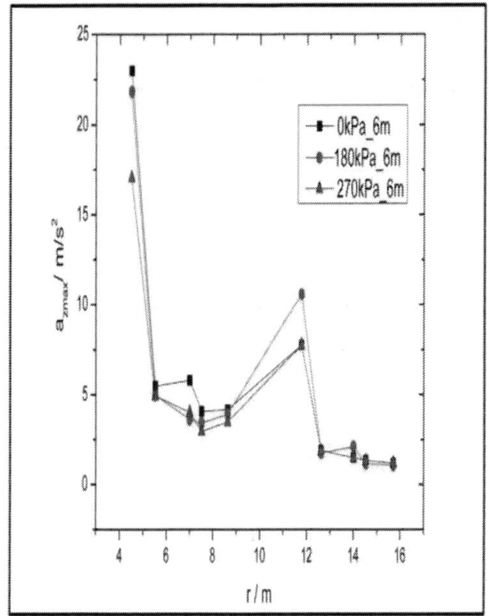

Figure 7. A_{zmax}-R Of CFGP Composite Foundation.

When the depth of the vibration source is 6 m and r = 7 - 8.6 m, with the no-load, the horizontal acceleration peak of the CFG pile composite foundation surface is at 2.44 - 1.53, and the vertical acceleration peak is 5.8 - 4.2; while pilling load is 270 kPa, the horizontal acceleration peak is 1.00 - 1.17, and vertical acceleration peak is 4.01 - 3.96.

When the distance from vibration source to the center of CFGP composite foundation distance is 14 m and the depth of vibration source is 6 m from the ground, the ratio of horizontal acceleration and vertical acceleration peak is at 0.61 - 2.75, the measuring point outside CFGP composite foundation when the distance r is 1 m, the ratio was 1.03, and near to 1.0. When the distance from vibration source to the center of CFGP composite foundation distance is 7 m and the depth of vibration source is 6 m from the ground, the ratio of horizontal acceleration and vertical acceleration peak is at 0.36 - 1.65; the farther the horizontal distance r from the measuring point to vibration source is, the greater the ratio.

Figures 8 and 9 are the peak acceleration with the horizontal distance r from the measuring point to the vibration source without no pilling load on the CFGLP-CFGSP composite foundation, when the depth of the vibration source is 6 m and the explosive quantity is 1.05 kg. From Figures 8-10, the peak acceleration along the CFGLP on the CFGLP-CFGSP composite foundation is near to peak acceleration of the CFGSP.

When the depth of the vibration source is 6 m and r is at 7 - 8.6 m, with the no-load, the horizontal peak acceleration of the CFGLP-CFGSP composite foundation surface is at 9.05 - 1.53 m/s^2, and the vertical peak acceleration is 13.27 - 2.72 m/s^2; while r is at 14 - 15.7 m, the horizontal peak acceleration is 5.23 - 2.25 m/s^2, and vertical peak acceleration is 5.13 - 0.7 m/s^2.

When the distance from vibration source to the center of CFGLP-CFGSP composite foundation distance is 14 m and the depth of vibration source is 6 m from the ground, the ratio of horizontal acceleration and vertical acceleration peak is at 1.02 - 3.15, measuring point distance from the vibration source is equal, the ratio is 1.49 on the CFGSP measuring point, and ratio is 3.15 on the CFGLP measuring point.

When the distance from vibration source to the center of CFGLP-CFGSP composite foundation distance is 7 m and the depth of vibration source is 6m from the ground, the ratio of horizontal acceleration and vertical acceleration peak is at 0.69 - 1.91, measuring point distance from the vibration source is equal, the ratio is 1.91 on the CFGSP measuring point, and ratio is 1.71 on the CFGLP measuring point. the farther the horizontal distance r from the measuring point to vibration source is, the greater the ratio.

Figures 11-18 are CFGLP and RCSSP peak acceleration changes of the CFGLP-RCSSP composite foundation with vibration source depth and pilling load, when the explosive quantity is 1.05 kg and pilling load is 360 kPa.
Figure 11 is the distribution law of the horizontal vibration peak acceleration on the CFGLP of the CFGLPRCSSP composite foundation. The depth of the vibration source is separately 2.5 m and 6 m, the horizontal vibration peak acceleration distribution is almost consistent. When the depth location of the vibration source is 7 m, the horizontal vibration peak acceleration maximum is at z = 4 m; when the location of the vibration source is 14 m, the horizontal vibration peak acceleration maximum is at z = 0 m.

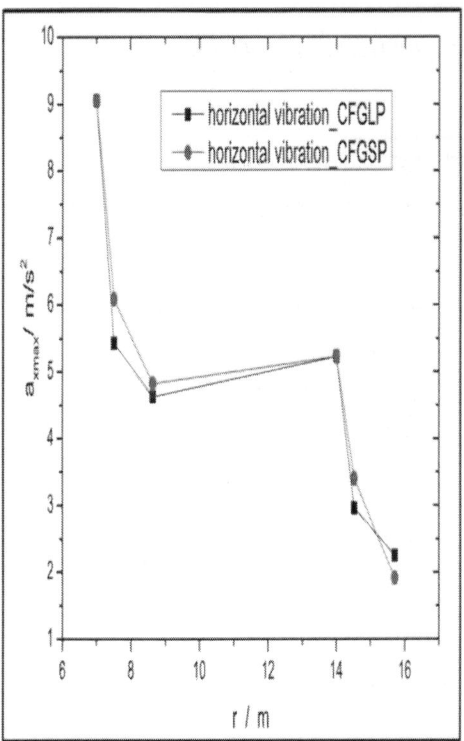

Figure 8. A_{xmax}-R of CFGLP-CFGSP Composite Foundation.

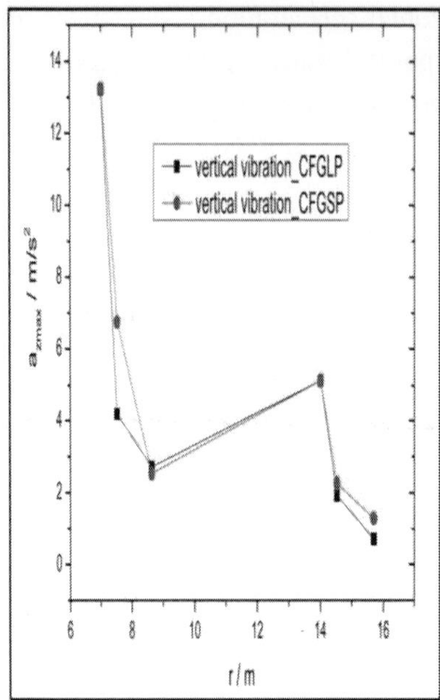

Figure 9. A_{zmax}-R Of CFGLP-CFGSP Composite Foundation,

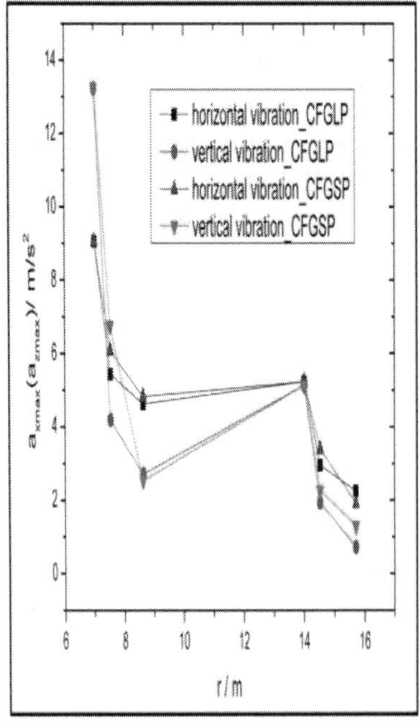

Figure 10. $A_{xmax}(A_{zmax})$-R Of CFGLP-CFGSP Composite Foundation.

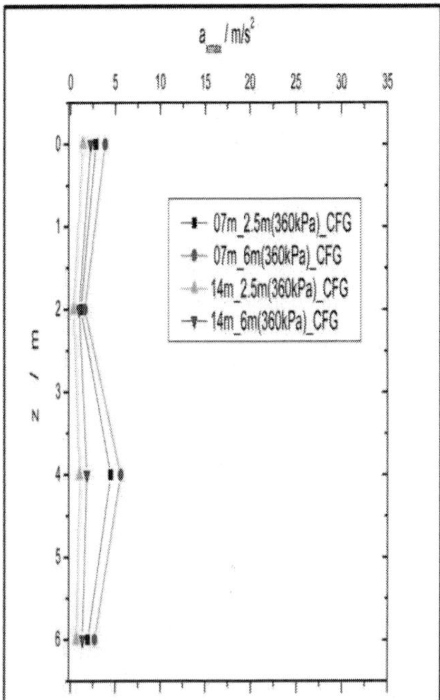

Figure 11. The Peak Acceleration A_{xmax} along the CFGLP,

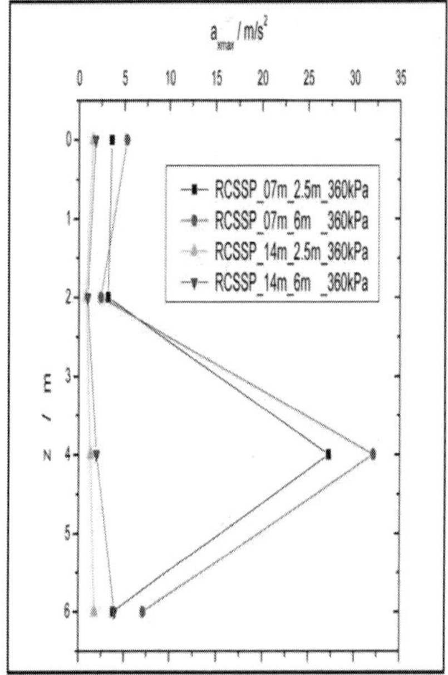

Figure 12. The Peak Acceleration A_{xmax} Along The RCSSP,

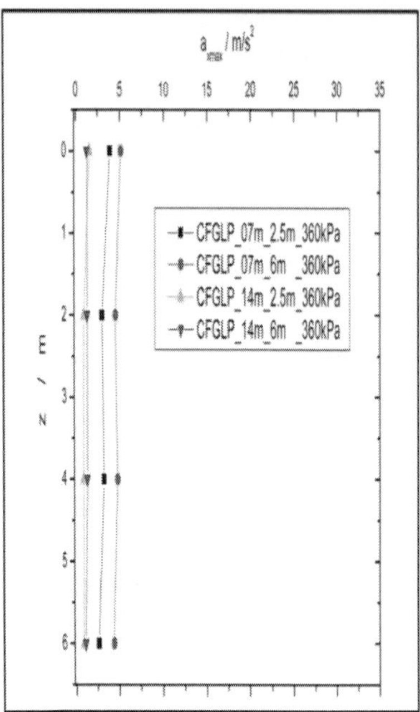

Figure 13. The Peak Acceleration A_{zmax} along the CFGLP.

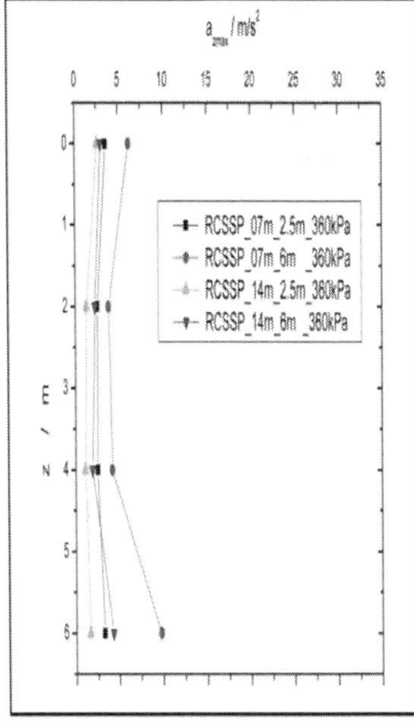

Figure 14. The Peak Acceleration A_{zmax} along the RCSSP,

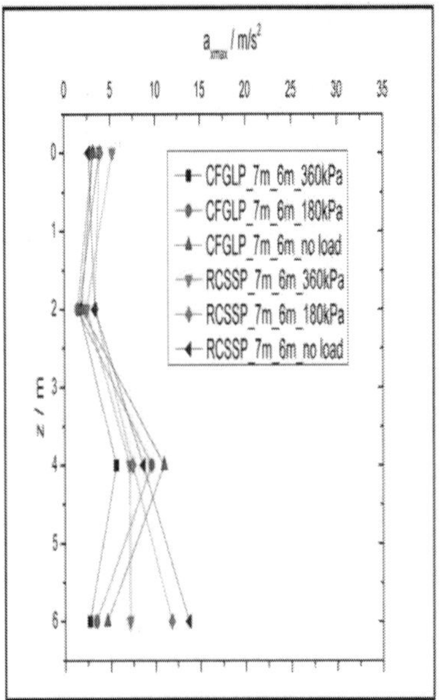

Figure 15. A_{xmax}-Z In CFGLP-RCSSP Composite Foundation,

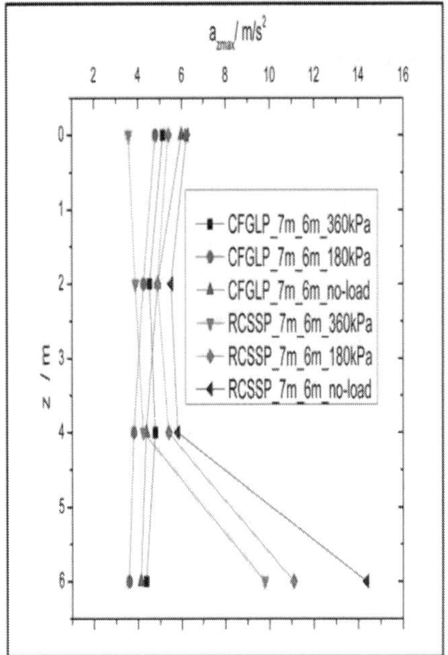

Figure 16. A_{zmax}-Z in CFGLP-RCSSP Composite Foundation,

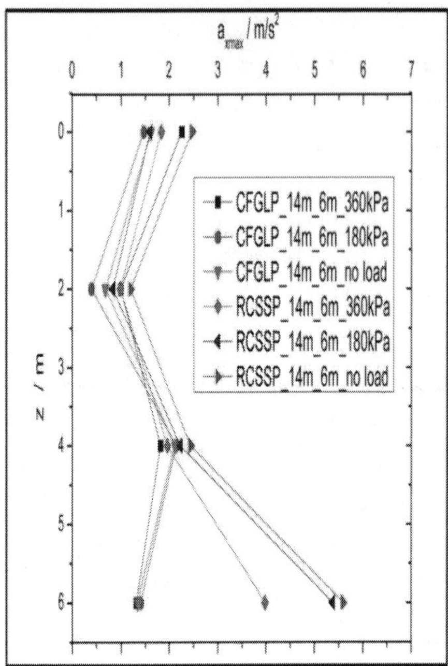

Figure 17. A_{xmax}-Z In CFGLP-RCSSP Composite Foundation,

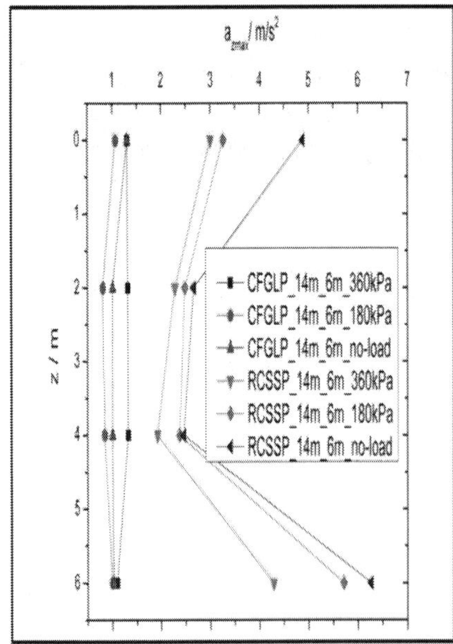

Figure 18. A_{zmax}-Z In CFGLP-RCSSP Composite Foundation,

Figure 12 is the distribution law of the horizontal vibration peak acceleration on the RCSSP of the CFGLPRCSSP composite foundation. When the depth location of the vibration source is 7 m, the horizontal vibration peak acceleration maximum is at $z = 4$ m; when the location of the vibration source is 14 m, the horizontal vibration peak acceleration maximum is at $z = 6$ m.

Figure 13 is the distribution law of the Vertical vibration peak acceleration on the CFGLP of the CFGLPRCSSP composite foundation. The vertical vibration acceleration peak on the CFGLP is change little, and the vertical vibration peak acceleration maximum is at $z = 0$ m.
Figure 14 is the distribution law of the Vertical vibration peak acceleration on the RCSSP of the CFGLPRCSSP composite foundation. The vertical vibration acceleration peak on the RCSSP is change little, and the vertical vibration peak acceleration maximum is at $z = 0$ m.

Figures 14-18 are the distribution law of the vibration peak acceleration along the pile of the CFGLP-RCSSP composite foundation. With the increase of pilling load on CFGLP-RCSSP of the composite foundation the peak acceleration is reduced, the change range of composite foundation of 4 m pile body acceleration is consistent, the peak acceleration of the measuring point of the RCSSP is larger than that of the CFGLP. The maximum of the horizontal acceleration peak occurs in $z = 4$ m or $z = 0$ m. The maximum of the vertical acceleration peak occurs in $z = 0$ m.

CONCLUSIONS

The nonlinear degree of the p - s curves of combined pile composite foundation decreases, and CFGLP-RCSSP is closed to linear relation. The bearing capacity of the four composite piles of the CFGP, CFGLP-CFGSP, and CFGLP-RCSSP in the site are separately 225 kPa, 179 kPa, and 197 kPa, separately increases 150%, 98.8% and 119% compared to the natural Foundation.

The field test shows that, under the same blast energy, vibration source position, form of composite foundation and properties of foundation soil, influence the main frequency of the composite foundation, but the rule is not obvious. The vibration main frequency is mainly depended on properties of foundation soil and piles between vibration source and measuring point, pilling load value. Horizontal vibration main frequency greater than the vertical vibration main frequency and the vertical vibration main frequency close to the first-order natural frequency of composite foundation. With the pilling load increasing, the CFGLP-RCSSP pile composite foundation combined frequency decreased.

The field test shows that, under the same blast energy, vibration source position, the acceleration peak on the CFGP composite foundation is less than CFGLP-CFGSP the corresponding values, as the load increases, the peak acceleration gently. CFGP composite foundation is favorable on seismic.

Field test shows that, under the same blast energy, vibration source positions, form of composite foundation, properties of foundation soil and pilling load have a significant effect on the peak acceleration of composite foundation. The distribution of peak acceleration is consistent within 4 m from pile top in the CFGLP-RCSSP composite foundation. The maximum of the horizontal acceleration peak along the pile body occurs at a distance of pile top 4 m or the pile top, and that of vertical acceleration peak occurred at a pile top.

REFERENCES

1.   (2002) GB50007-2002 Code for Design of Building Foundation(s). China Architecture Industry Press, Beijing.
2.   Yoshio, S. (1983) Deep Mixing Chemical Method Using Cement as Hardening Agent. Symposium on Soil and Rock Improvement Techniques, Bangkok, 34-37.
3.   Horii, N., Toyosawa, Y., Tamate, S. and Hashizume, H. (1998) Stability of Composite Ground Improved by Deep Mixing Method. Proceedings of 2nd International Conference on Ground Improvement Techniques, Singapore, 193- 198.
4.   Zheng, G. and Jiang, X.L. (1999) Research on the Bearing Capacity of Cement Treated Composite Foundation. Rock and Soil Mechanics, 3, 46-50.

5. Ma, H.L. (2003) Quantitative Analyses of the Influence of Pile Length and Other Factors on Capability and Modul of Cement Stabilized Soil Composite Foundation. Chinese Journal of Geotechnical Engineering, 11, 720-723.

6. Fu, J.H. and Song, E.X. (2000) Analysis of Rigid Pile Composite Foundation's Working Performance. Rock and Soil Mechanics, 21, 335-339.

7. Wang, M.-S., Wang, G.-C., Yan, X.-F., et al. (2005) In-Situ Tests on Bearing Behavior of Multi-Type-Pile Composite Subgrade. Chinese Journal of Geotechnical Engineering, 27, 1142-1145.

8. Yan, M.L., Wang, M.S., Yan, X.F. and Zhang, D.G. (2003) Study on the Calculation Method of Multi-Type-Pile Composite Foundation. Chinese Journal of Geotechnical Engineering, 25, 352-355.

9. Ding, J.H. (2007) Reliability Analysis on the Bearing Capacity of Composite Foundation with Multi-Type Compound Piles. Engineering Mechanics, 25, 168-172.

10. Ding, J.H., Liu, F.R. and Du, E.X. (2008) Dynamic Characteristic Analysis on Composite Foundation with Soil-Cement Piles and CFG Piles. Fly-Ash Comprehensive Utilization, 6, 37-40.

11. Wang, W.Y., Zhao, T. and Meng, Y.J. (2012) The Numerical Analysis on Dynamic Characteristics of CFG Pile Composite Foundation under Blasting. Engineering Mechanics, 29, 150-155.

12. Wang, W.Y., Zhao, T. and Ding, J.H. (2011) Effects on Dynamic Characteristics and Response of Rammed Soil-Cement Pile Composite Foundation. Engineering Mechanics, 28, 187-191.

13. Wang, W.Y., Zhao, T. and Ding, J.H. (2010) Influence Factors of Dynamic Characteristics and Response of CFG Pile Composite Foundation. Chinese Journal of Geotechnical Engineering, S2, 115-118.

14. Zhao, T., Yang, C.M. and Wang, W.Y. (2010) The Dynamic Test Research on CFG Pile Composite Foundation under Blasting. Journal of Highway and Transportation Research and Development (Applied Technique), 7, 121-122.

15. Ding, J.H., Wang, W.Y., Zhao, T., et al. (2013) The Dynamic Characteristic Experimental Method on the Composite Foundation with Rigid-Flexible Compound Piles. Open Journal of Civil Engineering, 3, 94-98. http://dx.doi.org/10.4236/ojce.2013.32010

CITATION

Ding, J., Cao, Y. , Wang, W. , Zhao, T. and Feng, J. (2014) Experimental Study of Dynamic Characteristics on Composite Foundation with CFG Long Pile and Rammed Cement-Soil Short Pile. *Open Journal of Civil Engineering*, **4**, 1-12. doi: 10.4236/ojce.2014.41001.

CHAPTER 4

Influence of the Soil-Structure Interaction in The Behavior of Mat Foundation

Oustasse Abdoulaye Sall[1], Meissa Fall[1], Yves Berthaud[2], Makhaly Ba[1], Mapathé Ndiaye[1]*

[1]Département Of Génie Civil, UFR SI-Université De Thiès, Thiès, Sénégal

[2]Université Pierre Et Marie Curie, Paris, France

INTRODUCTION

To calculate the displacement in a foundation most of civil engineering structure computing software admits a constant elastic modulus. The calculation of displacements in a foundation requires to model realistically the behavior of the soil foundation. This requires the definition of models of behavior adapted to foundation soils. To approach the reality, a linear elastic constitutive law, perfectly plastic (such as Mohr Coulomb law) can be introduced (Coquillay, 2005) [1] for varying linearly the parameters (E, v, c ⋯) with the depth. It is generally agreed that the Young's modulus of homogeneous soil increases with depth. This can be explained by an increase in the mean stress and the densification it leads. In this context ,the variation of the soil elastic modulus (E_s) can be given by the following expression:

$$E_s = E_1 \left(1 - z/H\right) + \frac{z}{H} E_2$$

$$(1)$$

where E_1 and E_2 are the elastic modulus respectively at the top and the bottom of the soil mass.

In what follows, the profile of displacements in the soil mass is studied. To do this it is assumed at first that $\phi(z)$ and the elastic modulus (E_s) of the soil foundation are linear, and the Poisson ratio of the soil (v_s) is constant with the depth of the foundation massif. Secondly, the linearity of Poisson's ratio and an hyperbolic sine variation of the function $\phi(z)$ are introduced.

MODELISATION

Using the model of the elastic continuum, Vlazov (1949) [2] proposed a formulation for modeling soil, based on the application of the variational method. The soil layer of thickness H is considered as a linear isotropic elastic medium, resting on a rigid substrate (Figure 1) [3] . The horizontal displacements u(x, y, z) and v(x, y, z) are assumed to be zero in the whole mass of soil. The vertical displacements are given by:

$$w(x, y, z) = w(x, y)\phi(z) \tag{2}$$

where $\phi(z)$ is a function which describes the variation of the displacements w(x, y) along the z axis, such that: $\phi(0) = 1$ and $\phi(H) = 0$.

$\phi(z)$ can be expressed by the following relationships:

$$\phi(z) = (1 - z/H) \tag{3}$$

$$\phi(Z) = \frac{\sinh\left[(H - z)\dfrac{\gamma}{L}\right]}{\sinh\left(\dfrac{\gamma H}{L}\right)} \tag{4}$$

γ and L are constants parameters.

The equilibrium equations in the z direction are obtained by applying the principle of virtual work (Turhan, 1992) [4] . Thus the total response of the system under a load (q), taking into account the soil-structure interaction, is given by:

$$D\left(\frac{\partial^4 w}{\partial x^4} + 2\frac{\partial^4 w}{\partial x^2 \partial y^2} + \frac{\partial^4 w}{\partial y^4}\right) - 2T\left(\frac{\partial^2 w}{\partial x^2} + \frac{\partial^2 w}{\partial y^2}\right) + kw = q(x,y)$$

(5)

where D is the flexural rigidity of the plate and is given by:

$$D = \frac{E_b e^3}{12\left(1 - v_b^2\right)}$$

(6)

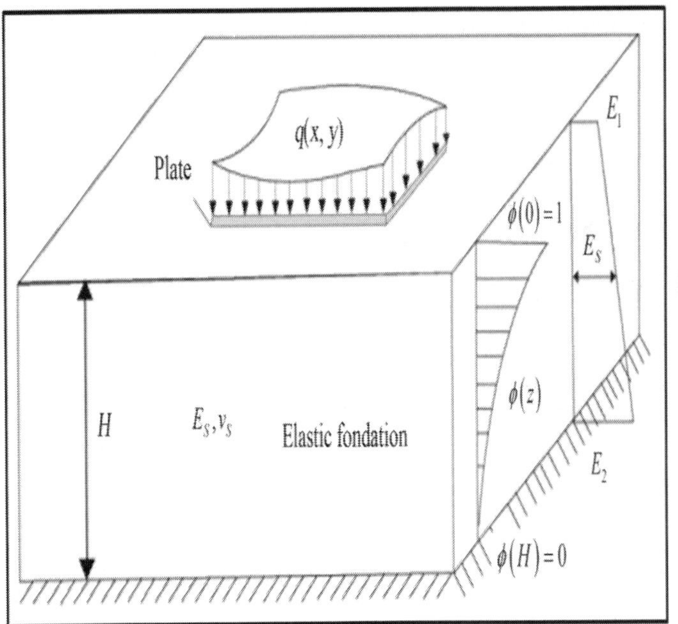

Figure 1. Vlasov model, stresses in an elastic layer (Selvadurai, 79).

with: E_b: elastic modulus of the platee: the thickness of the platev_b: the Poisson's ratio of the plate.2T and k are the two parameters of the Vlasov model;k is the modulus of subgrade reaction expressed by Biot (1935) [5] according to the following relationship:

$$k = \frac{0.65E_s}{1-v_s^2} \times 12\sqrt{\frac{E_s B^4}{E_b I}}$$

(7)

Vesic (1963) [6] improved the formula of Biot and proposed the following relationship:

$$k = \frac{0.95E_s}{1-v_s^2}\left(\frac{E_s B^4}{\left(1-v_s^2\right)E_b I}\right)^{0.108}$$

(8)

where:

E_s is the modulus of subgrade;

v_s is the Poisson's ratio of subgrade;

B is the width of the foundation;

E_b is the Young's modulus of the concrete;

I is the moment of inertia of the cross section of the concrete.

T is the horizontal elastic modulus of subgrade reaction proposed by Vlasov (1949) as follows:

$$T = \frac{E_s}{4\left(1-v_s^2\right)\left(1+v_s\left(1-v_s\right)\right)}\int_0^H \phi^2 dz$$

(9)

For a relatively deep layer of soil where the normal stresses may vary with depth, it is possible to use, for the function $\phi(z)$, the non-linear continuously variation defined by Equation (4).

Indeed, the profile of hyperbolic sine requires an estimation of the arbitrary parameter γ which Vlazov (1949) [2] did not specify the value. This author just recommended values between 1 and 2.

ANALYTICAL SOLUTION

For the case of an elastic homogeneous soil, it can be assumed a uniform distribution of the applied force on the foundation system. This amount to an admission that the stress q(x, y) is a constant Q value in every point of the structure. For a foundation of infinite dimension, zero displacement at the edges of the plate is imposed. Using the double Fourier series and accounting for the displacement of the soil under the effect of the own weight of the slab, the displacement at any point of the system is given by:

$$w(x,y,z) = \frac{16Q}{\pi^2}\left(\sum_{1,3,5}^{\infty}\sum_{1,3,5}^{\infty}\frac{\sin\left(\frac{m\pi x}{L}\right)\times\sin\left(\frac{n\pi y}{B}\right)}{Dmn\left(\left(\frac{m\pi}{L}\right)^2+\left(\frac{n\pi}{B}\right)^2\right)^2+2Tmn\left(\left(\frac{m\pi}{L}\right)^2+\left(\frac{n\pi}{B}\right)^2\right)+kmn}+25000\times\frac{e}{k}\right)\times\phi(z)$$

(10)

The maximum values of m and n are set to 45 as above 25 the sensibility relative to m and n is no longer perceptible.

It appears from the study of Sall et al. (2013) [7] that the elastic modulus and the Poisson's ratio of the subgrade are the most influential parameters on the displacements of the plate. The modulus of subgrade reaction and displacements vary slightly with the mechanical properties of concrete and are more influenced by the elastic modulus of the soil.

NUMERICAL VALIDATION OF RESULTS

In order to validate the results, a concrete slab resting on an elastic soil with constant Young's modulus is considered.

With B = 20 m, L = 20 m, e = 0.4 m; E_b = 33 GPa, E_s = 7 MPa, v_s = 0.2 and Q = 15 t/m^2.

The values of the vertical according to Biot and horizontal elastic modulus are calculated and fed into a finite element software ("Flexion des plaques" under "RDM 6" module) leads to the results. After analyzing the results from "Flexion des plaques" the Castem[6] finite element code is used to understand the influence of parameters on the results. k values given by Biot were used because they are lower than those given by Vesic. The more k is significant, the more displacements are low. However, to avoid underestimating displacements, k values given by Biot are used in this research.

Analysis of Displacements Provided by "Flexion Des Plaques"

The mesh is given by the Figure 2.

For boundary conditions, it is assumed a displacement equal to the displacement of the interface under the effect of the weight of the slab.

The results are given in Figure 3 which shows that the displacement is most important at the center of the plate.

The following figure visualizes the shape of the displacement along the median of the plate. From Figure 4, it can be said that there is a correlation between the displacements given by the analytical model and those of the finite element method ("Flexion des plaques").

Note that the maximum displacement obtained in the numerical calculation is 5 mm higher than the maximum value of the analytical calculation. This difference in displacement values can be explained by the fact that during the numerical simulation, the elastic supports match the nodes corresponding to the mesh.

But in reality, the elastic support is extended under all the plate which contributes to reduce the displacements.

For the same given parameters, the influence of Poisson's ratio of concrete on the displacements is evaluated. Table 1 gives a summary of this influence on the maximum displacement (w_{max}).

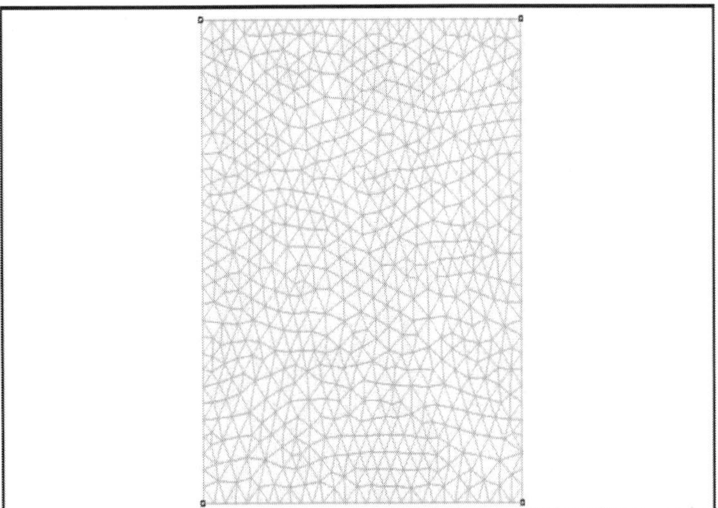

Figure 2. Mesh of the plate by finite elements.

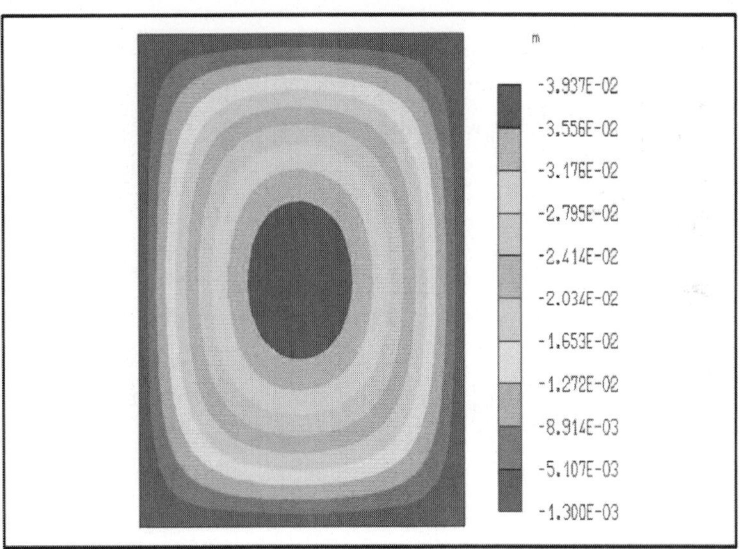

Figure 3. Displacements Cartography.

Table 1. W_{max} According To N_b.

(v_b)	0.2	0.25	0.3	0.4	0.45
w_{max} (cm)	3.753	3.752	3.75	3.744	3.74

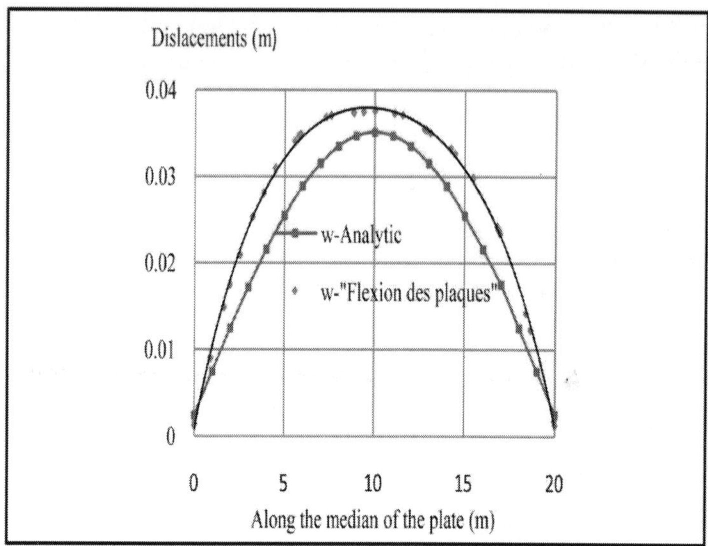

Figure 4. Displacements along the Median of the Plate.

Table 1 shows a weak influence of Poisson's ratio on the displacements of the soil-structure interface. These displacements vary in the range of one to four hundredths of a millimeter when the Poisson's ratio ranges from 0.25 to 0.45. Then, the displacements are quasi-independent of Poisson's ratio of the concrete.

In the following, it is proposed to study the influence of the elastic modulus of the concrete foundation displacements of the structure as shown in Table 2. To do so, a concrete slab of the same dimensions (20 m × 20 m × 40 cm) with a Young's modulus ranging between 33 and 46 GPa, is subjected to an uniform loading (15 t/m²). The slab rests on an elastic soil of constant Young's modulus equal to 7 MPa and Poisson's ratio equal 0.2. The different values of w_{max} for various values of the elastic modulus of the concrete are shown in Table 2. TheFigures 5-7 show the displacements cartography for various values of E_b.

Table 2 gives a summary of the maximum value for different Young's modulus of concrete.

The above analysis shows that the Young's modulus of the concrete has an insignificant influence on the displacement of the plate resting on elastic foundation.

To study the influence of the elastic modulus of the subgrade on the displacements of the plate, the previous data are considered, the Young's modulus of concrete is set as at a constant value and the modulus of the soil is varying from 4 to 8 MPa.

It appears from this study that the elastic modulus of subgrade has a considerable influence on the displacements of the plate (Figures 8-10).

Analysis of Displacements Provided by Castem

In order to confirm the validity of the results, another finite element code (Castem[ó]) is used to study the influence of different model parameters. The results provided by Castem[ó] will be compared with results provided by the analytical model and reported in the Figure 11.

The influence of other parameters provided by Castem[ó] is given in Tables 3-6:

It is clear from this analysis that the parameters of the subgrade (E_s, v_s) have a greatest influence on the displacements of the plate. The properties of the concrete (E_b, v_b) have a weak influence on the displacements. This which just comfort the analysis of analytical results given by Sall et al. [7].

In what follows, the influence of mechanical properties of soil and concrete foundation on displacements of soil mass is studied.

Figures 12-17 show the influence of model parameters on displacements of the soil foundation. These results show that soil parameters are more influential than those of the concrete foundation. It should be noted that the concrete properties, especially the Poisson's ratio, have not substantially influenced the displacements profile (Figure 16).

These results show that the elastic modulus of the substrate has an influence more or less significant on the displacements of points located above the middle of the soil mass and a less pronounced effect on the displacements of points near the bedrock and the top of the soil mass (Figure 13).

Table 2. W_{max} Values According To N_b.

v_b (GPa)	43	36	33
w_{max} (10^{-2} m)	2.99	3.02	3.082

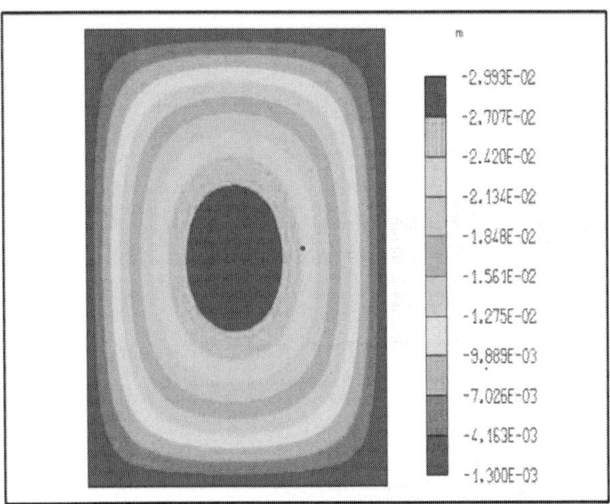

Figure 5. Displacements cartography for $E_b = 46$ Gpa.

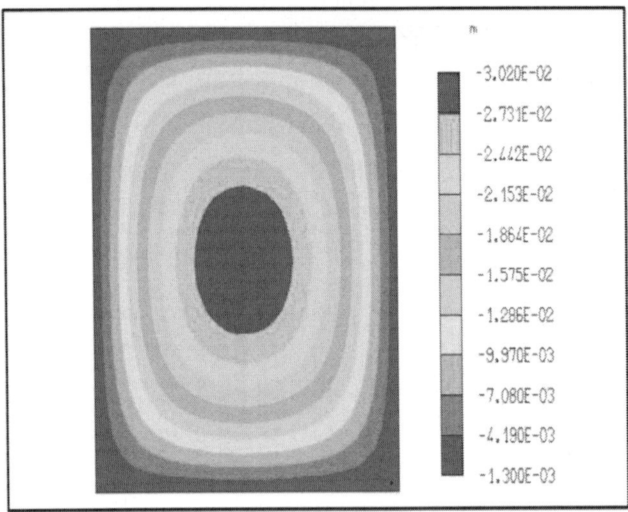

Figure 6. Displacements cartography for $E_b = 36$ Gpa.

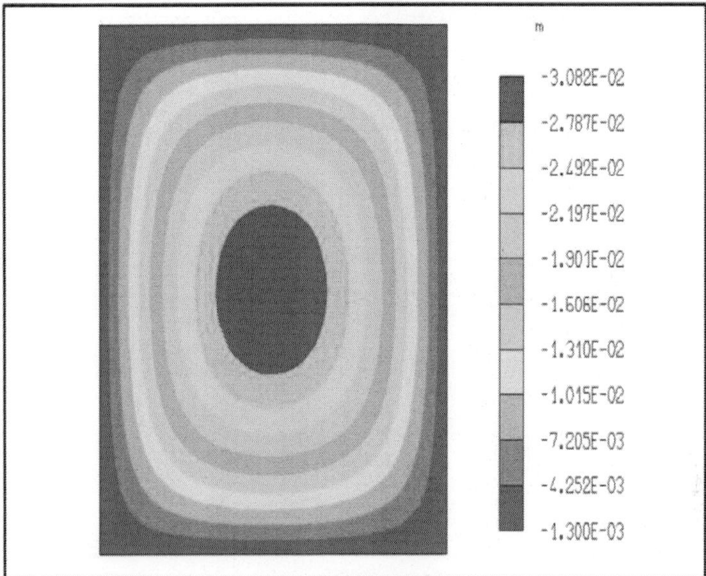

Figure 7. Displacements cartography for E_b = 33 Gpa.

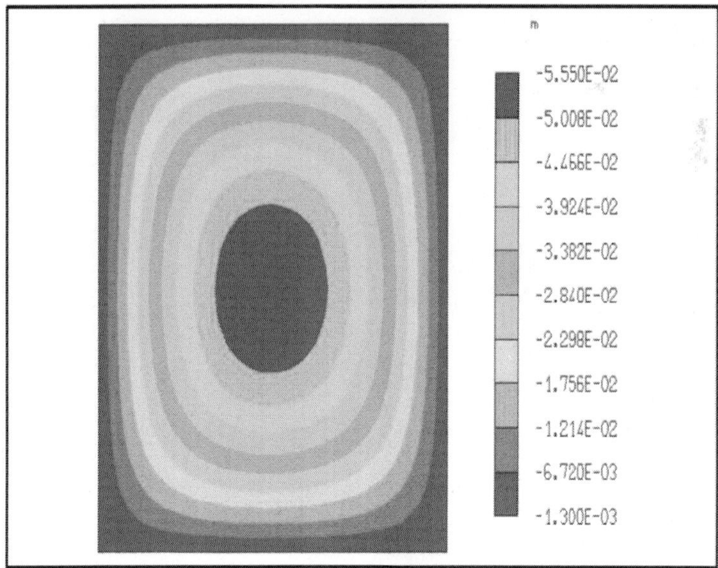

Figure 8. Displacements cartography for E_s = 4 Mpa.

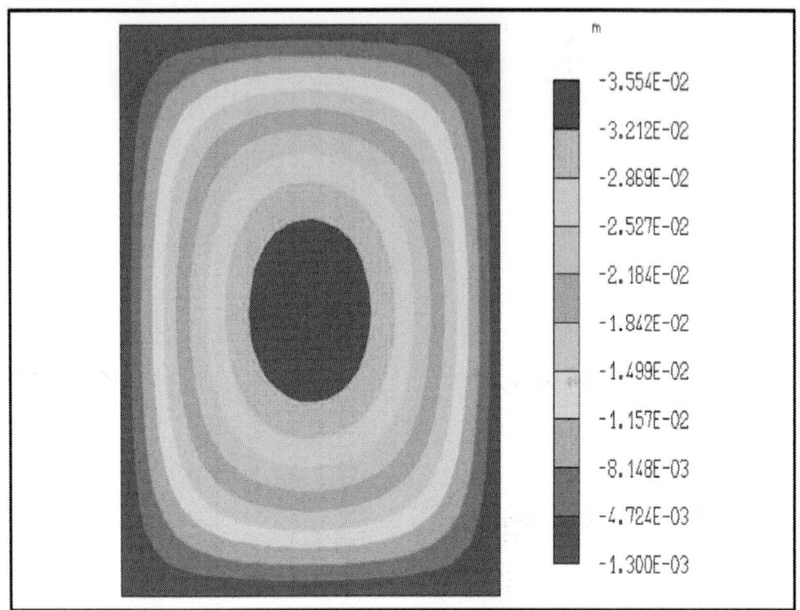

Figure 9. Displacements cartography for $E_s = 6$ Mpa.

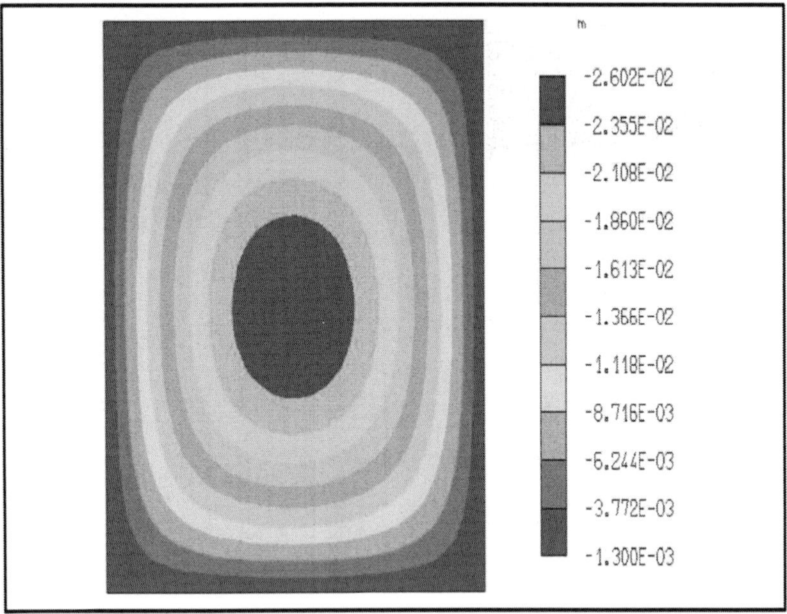

Figure 10. Displacements cartography for $E_s = 8$ Mpa.

Table 3. W_{max} Values According Castem[6] According To E_b.

E_b (GPa)	33	36	39	43
$w_{max.}$ (cm)	4.01793	4.01792	4.01791	4.01790

Table 4. W_{max} Values According Castem[6] According To V_s.

V_s	0.2	0.25	0.35	0.4
$w_{max.}$ (cm)	4.01793	3.72031	2.78166	2.0834

Table 5. W_{max} Values According Castem[6] According To V_b.

V_b	0.2	0.25	0.35	0.4
$w_{max.}$ (cm)	4.01793	4.01792	4.01790	4.01789

Table 6. W_{max} Values According Castem[6] According To E.

e (cm)	20	40	60	80
$w_{max.}$ (cm)	4.01789	4.01793	4.01796	4.01799

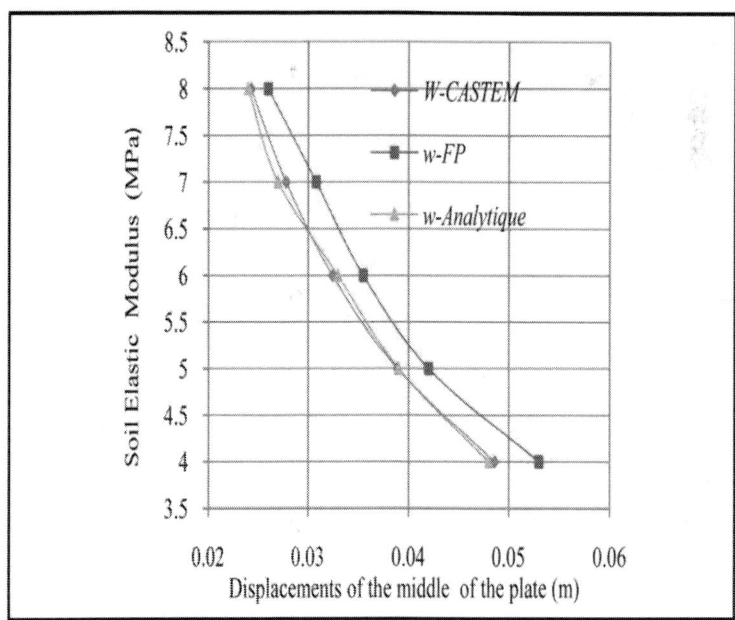

Figure 11. Displacements of the middle of the plate according to E_s,

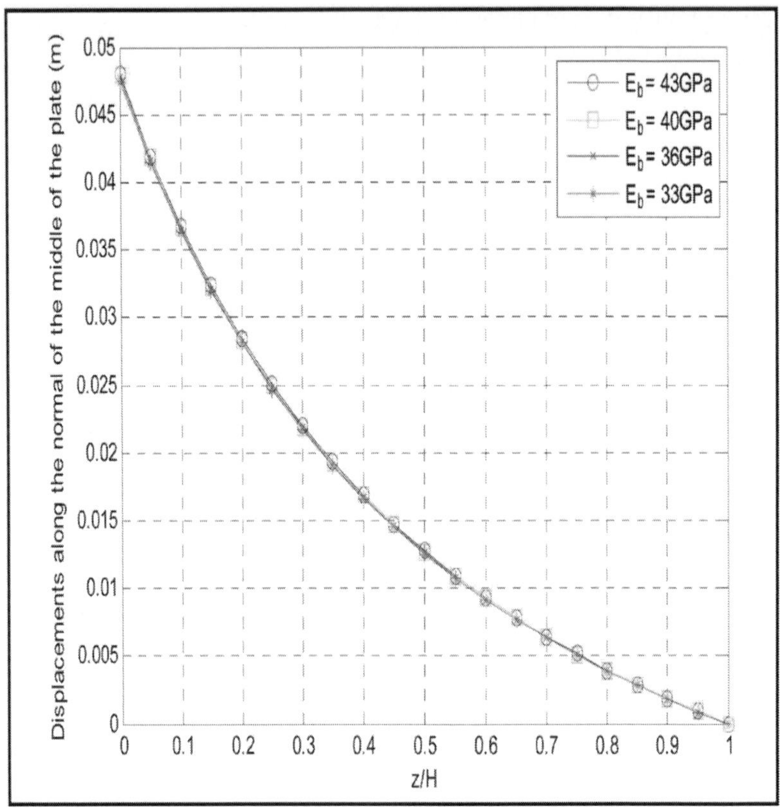

Figure 12. Profile of the displacements from the interface to the rigid substratum for various values of E_b.

These results show also a discernible influence of the elastic modulus of the top of the soil. The influence of Young's modulus of top of the soil is more pronounced on the half-height above the soil. The upper half-height is the most affected part of the soil. Figure 18 shows the influence of the ratio B/L on the profile of displacements in the soil.

In which follows in addition to the linearity of the soil elastic modulus, the linearity of Poisson's ratio is introduced. It is assumed that variation of $\phi(z)$ follows a hyperbolic sine. The Poisson's ratio in this context is given by:

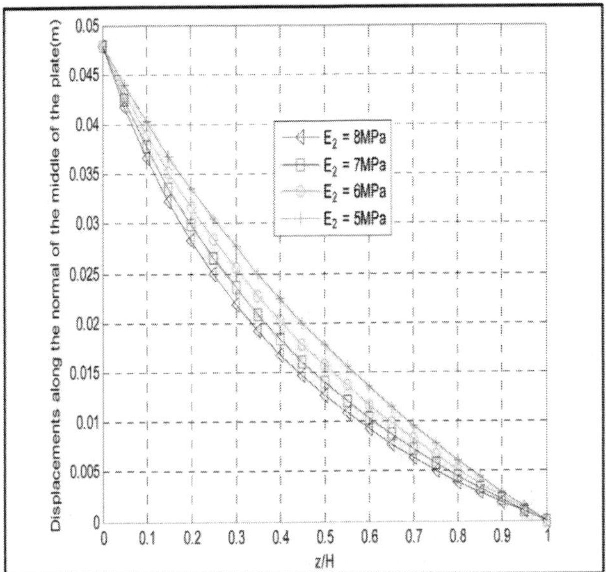

Figure 13. Profile of the displacements from the interface to the rigid substratum for various values of E_2.

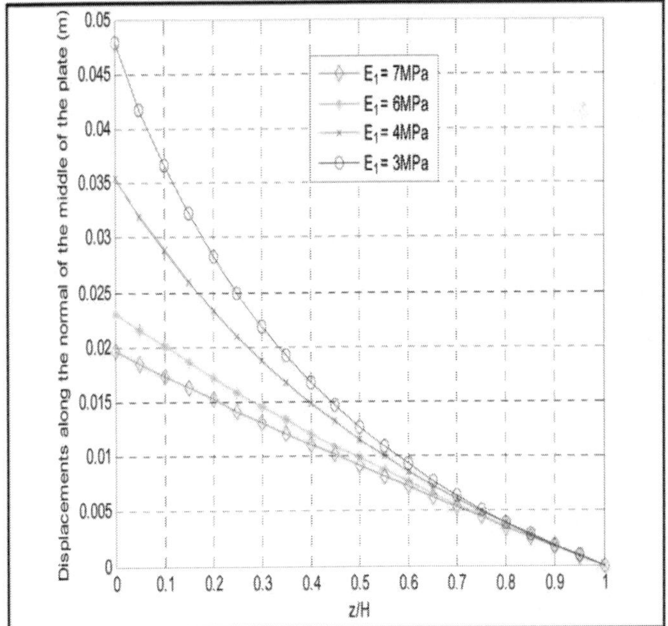

Figure 14. Profile of the displacements from the interface to the rigid substratum for various values of E_1.

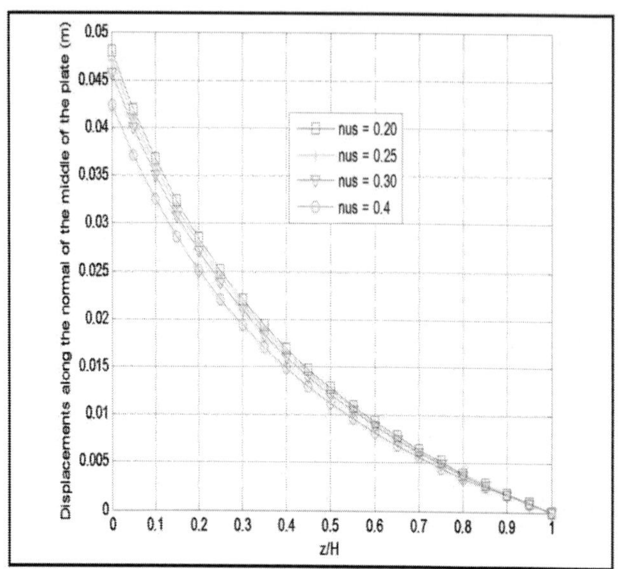

Figure 15. Profile of the displacements from the interface to the rigid substratum for various values of V_s.

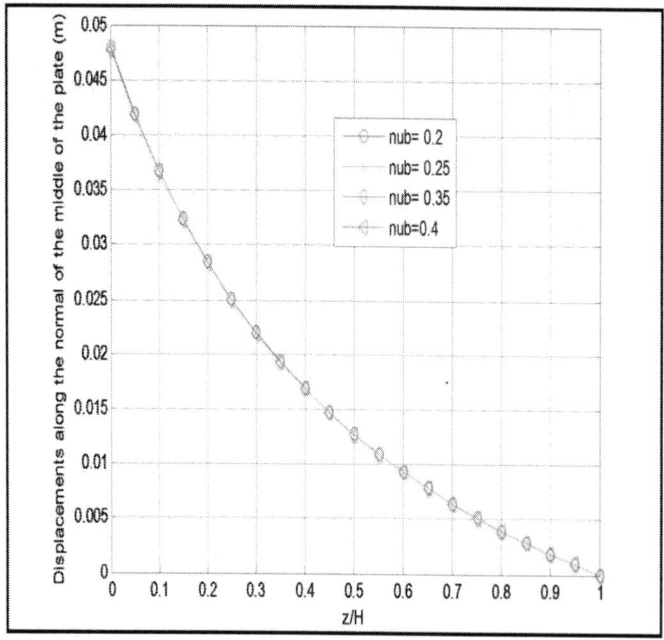

Figure 16. Profile of the displacements from the interface to the rigid substratum for various values of N_b.

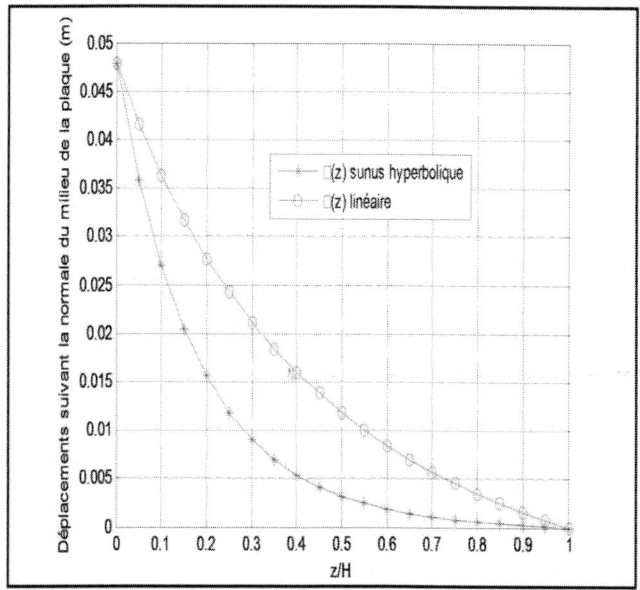

Figure 17. Profile of the displacements from the interface to the rigid substratum for various expression of $\Phi(Z)$.

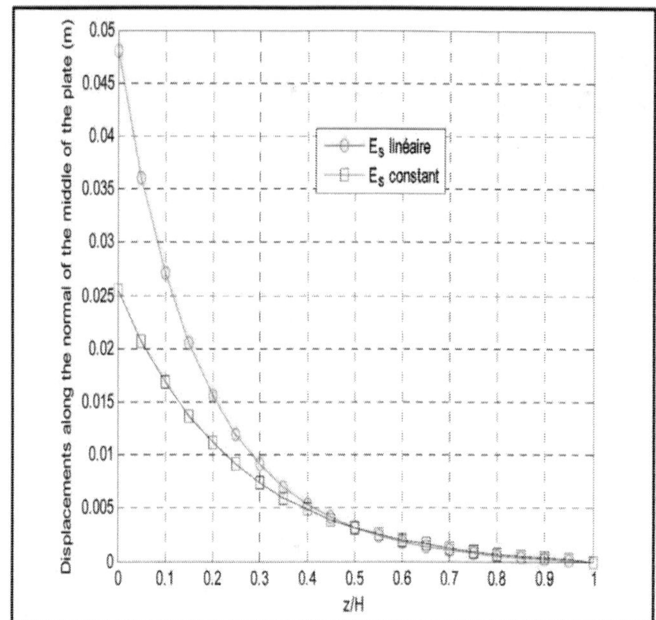

Figure 18. Profile of the displacements from the interface to the rigid substratum for linear and constant E_s,

$$v_s = v_1\left(1 - z/H\right) + \frac{z}{H}v_2$$

(11)

with v_1 and v_2 respectively representing the Poisson's ratios of the top and the bottom of the soil.

Figures 17-24 show the influence of the parameters of a linear model with the elastic modulus and Poisson's ratio. It is clear from this analysis that the settlements in the soil mass are more pronounced for $\phi(z)$ hyperbolic sine for a linear profile (Figure 17).

Figure 17 shows the importance of controlling the stress path and strain in the soil foundation. The linearity of the elastic modulus of soil gives more pronounced displacements in the soil (especially on the upper half thick of the soil) than when a constant elastic modulus of subgrade is admitted.

These results also show a small influence of the linearity of Poisson's ratio on the displacements profile (Figure 22).

Figure 23 shows the influence of the thickness on the profile of the settlements in the soil. The influence of the thickness of the plate is more observed on the upper half thickness of the soil.

This figure confirms that the thicker is the plate the more important are settlements in the soil massif.

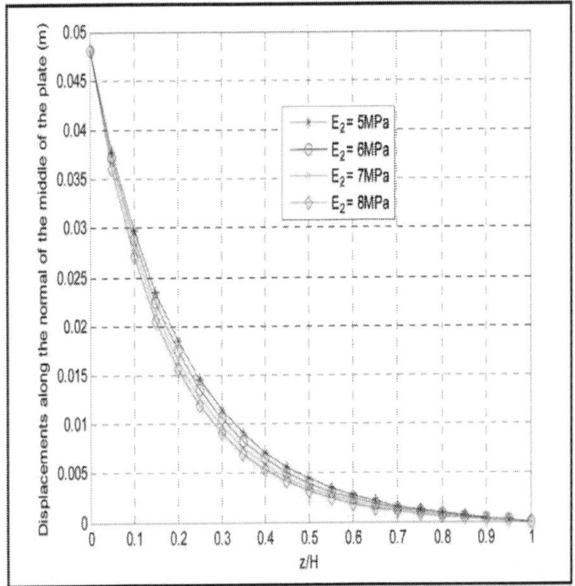

Figure 19. Profile of the Displacements from The Interface To The Rigid Substratum For Various Values Of E_2.

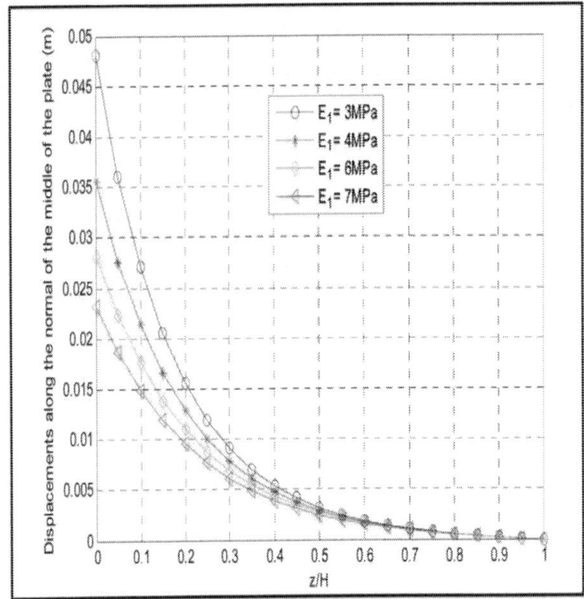

Figure 20. Profile of the Displacements from the Interface to the Rigid Substratum For Various Values Of E_1.

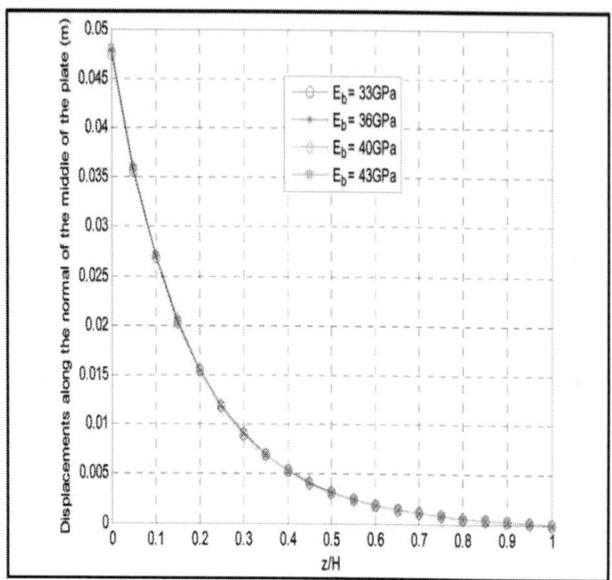

Figure 21. Profile of the Displacements from the Interface to the Rigid Substratum For Various Values Of E_b.

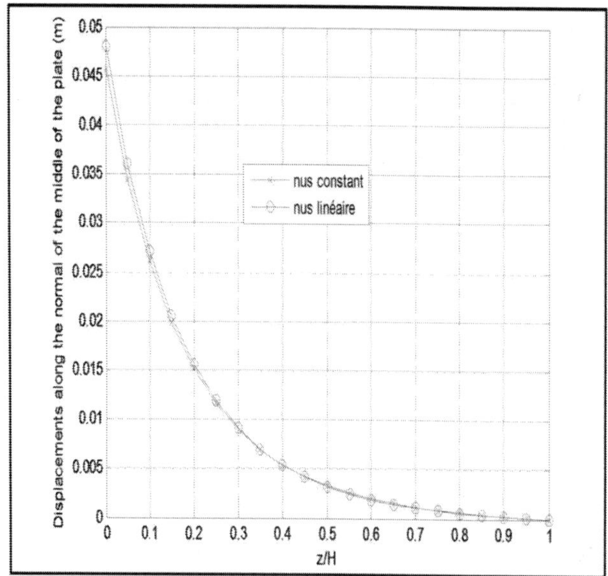

Figure 22. Profile of the Displacements From The Interface To The Rigid Substratum For Linear And Constant V_s.

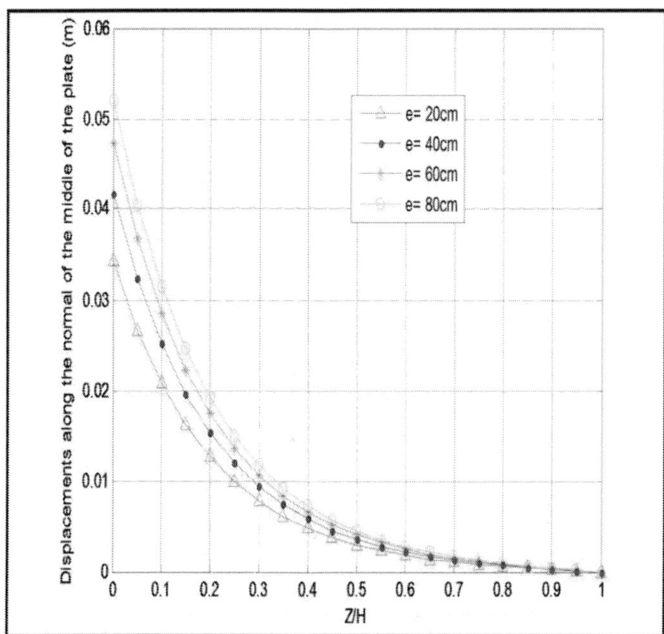

Figure 23. Profile of the displacements from the interface to the rigid substratum for various values of E.

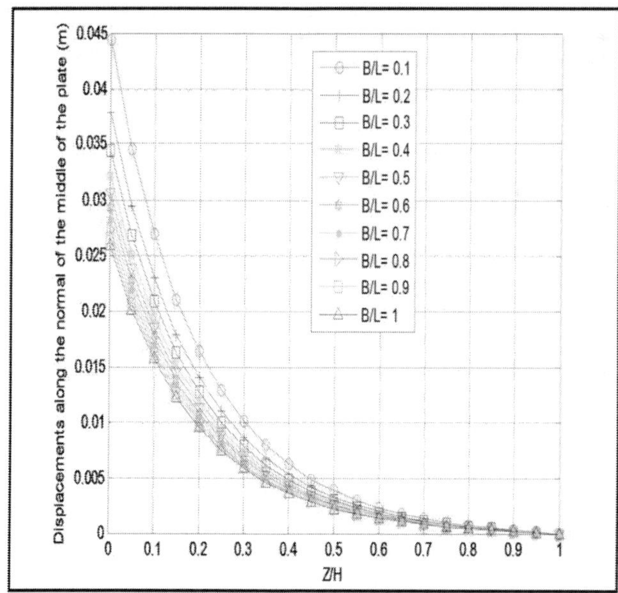

Figure 24. Profile of the displacements from the interface to the rigid substratum for various values of B/L,

It is clear from these findings that the influence of model parameters on the profile of displacements is more pronounced on the upper half-thickness of the soil.

Figure 18 shows the displacements for constant and linear soil modulus. It is clear from this figure that taking a constant modulus leads to underestimate displacements because linearity of E_s gives greater displacements at points on the upper half-thickness of the soil.

CONCLUSION

It appears from this study that the linearity of the elastic modulus of subgrade leads to greater displacement than when this modulus is assumed to be constant in the soil. The points located in the upper half-thickness of soil are the most sensitive to the parameters of the model. Poisson's ratio has an insignificant influence on the observed displacements. This study shows that the mastery of mechanical properties and the path of displacements in the soil is important for mastering the future life of the structure where the need to pay utmost importance on soil studies before calculation and construction of any foundation. This study shows that the concrete alone does not solve all problems related to disorders in a foundation and mastery of soil parameters is important to minimize disorders.

REFERENCES

1. Biot, M.A. (1937) Bending of an Infinite Beam on an Elastic Foundation. Journal of Applied Physics, 12, 155-164. http://dx.doi.org/10.1063/1.1712886
2. Coquillay, S. (2005) Prise en Compte de la non Linéarité du Comportement des sols Soumis à de Petites Déformations pour le Calcul des Ouvrages Géotechniques. Thèse de Doctorat de l'Ecole Nationale des Ponts et Chaussées, Paris.
3. Selvadurai, A.P.S. (1979) Elastic Analysis of Soil-Foundation Interaction. Developments in Geotechnical Engineering, 17, Elsevier Scientific Publishing Company, New York.

4. Turhan, A. (1992) A Consistent Vlasov Model for Analysis on Plates on Elastic Foundation Using the Finite Element Method. The Graduate Faculty of Texas Tech University in Partial Fulfillment of the Requirements for the Degree of Doctor.
5. Vesic, A.B. (1963) Beams on Elastic Subgrade and the Winkler's Hypothesis. Proceedings of the 5th International Conference of Soil Mechanics, 845-850.
6. Vlazov, V.Z. and Leontiev, U.N. (1966) Beams, Plates and Shells on Elastic Foundations. Israel Program for Scientific Translations, Jerusalem.

CITATION

Sall, O, Fall, M , Berthaud, Y. , Ba, M. and Ndiaye, M. (2014) Influence of the Soil-Structure Interaction in the Behavior of Mat Foundation. *Open Journal of Civil Engineering*, **4**, 71-83. doi: 10.4236/ojce.2014.41007.

CHAPTER 5

Assessment of Seismic Vulnerability of a Historical Masonry Building

Francesca Ceroni, Marisa Pecce, Stefania Sica and Angelo Garofano

Department of Engineering, University of Sannio, Piazza Roma, 21-82100 Benevento, Italy

ABSTRACT

A multidisciplinary approach for assessing the seismic vulnerability of heritage masonry buildings is described throughout the paper. The procedure is applied to a specific case study that represents a very common typology of masonry building in Italy. The seismic vulnerability of the examined building was assessed after the following: (a) historical investigation about the building and the surrounding area, (b) detailed geometrical relieves, (c) identification of materials by means of surveys and literature indications, (d) dynamic *in-situ* tests, (e) foundation soil characterization, (f) dynamic identification of the structure by means of a refined Finite Element (FE) model. After these steps, the FE model was used to assess the safety level of the building by means of non-linear static analyses according to the provisions of Eurocode 8 and estimate of the q-factor. Some parametric studies were also carried out by means of both linear dynamic and non-linear static analyses.

KEYWORDS

Masonry buildings; Knowledge level; Multidisciplinary *in-situ* inquiries; FE models; Dynamic behavior; Non-linear analyses

INTRODUCTION

The assessment of safety and, in general, of the structural behavior of existing masonry buildings is often a complex issue due to the inevitable uncertainness regarding geometry, typologies and mechanical properties of materials, effects of age and past loading history referring both to service and accidental loads (earthquake, fire, explosion, *etc.*), presence of architectonical and structural interventions with different techniques or materials, interaction with surrounding buildings and foundations, knowledge of foundation soil properties. Thus, for an existing masonry building, a complete knowledge is needed before starting any study of its structural safety or designing an eventual retrofitting intervention.

A complete knowledge needs detailed *in situ* inquiries according to a multidisciplinary approach providing both traditional and innovative investigation techniques. A detailed knowledge of the structure, even if it could appear expensive and time-consuming in the initial stage, in the later stages allows optimizing an eventual repairing design; such an approach reduces the costs of intervention and is in agreement with the 'minimum intervention' philosophy for heritage buildings [1,2]. Indeed, the knowledge level of the geometrical and structural characteristics of the examined building strictly influences the definition of the numerical models and, thus, the reliability of the building response prediction. The Italian Guidelines for cultural heritage [2] suggest to use methods characterized by different levels of knowledge and detail for the assessment of the seismic safety of historical buildings in function of the aim of such studies. The final level of knowledge of the structure depends on quality, quantity and deepening of *in situ* relieves, historical investigations, and experimental tests. The knowledge phases are aimed to define a model able to predict qualitatively the structural behavior or to carry out more refined structural analyses. Clearly, the reliability of each type of model, qualitative or quantitative, depends on the deepening level of the inquiries and on the amount of available data.

Simple approaches, based on indexes related both to a simplified estimation of the seismic resistance of the structure and the expected seismic action [2], can be useful instruments to develop seismic vulnerability maps functional to define a priority scale for further detailed analysis.

More refined analyses request a larger and more detailed database of information about the examined structure (geometry, mechanical properties in the linear and non-linear field, variability of parameters and anomalies in the constituent elements, *etc.*....) that can be collected by several fonts. Furthermore, for historical buildings the simplified methods can give useful indications particularly when detailed experimental *in situ* investigations cannot be carried out, *i.e.*, in presence of strong preservation requirements, so that detailed approaches become unreliable. Note that the simplified methods are in general reliable only for regular and symmetrical buildings, stiff floors and predominant shear failure modes in the walls.

Historical investigations can be useful to characterize and analyze the origin of the structure and its vicissitudes. In particular, the following topics can be assessed by means of this type of investigation:

- the reasons for earlier structural interventions: severe damages or collapse due to catastrophic events (earthquake, explosion, fire) or normal use of the building, change of use destination, change of owners, change of architectonic features related to the current architectonic style;
- the overall evolution of the examined building as well as of the surrounding buildings related to the urban evolution of the area;
- the presence of not contemporaneous construction phases characterized by different materials and techniques;
- the individuation in coeval buildings placed in the same area of typical construction materials and techniques. Indeed, the choice of the construction materials, especially if natural carved stones are used, was often related to the geographical location of the building and the local availability of quarries in the neighborhood.

Most of this information can be collected only by means of archival and bibliographical documentations and their importance should be not neglected.

Successively, detailed surveys of geometry and materials make it possible to detect characteristic features such as type and configuration of floors, disposition of resistant masonry walls, nature and dimensions of masonry component elements (*i.e.*, type and dimensions of masonry blocks), wall texture (*i.e.*, disposition of blocks and mortar joints, thickness of joints, presence of a single or double panel with chopped materials in the central cavity), construction details (connections between the floors and the vertical resistant elements, local connections between parallel walls as steel chains or between orthogonal walls in the corners along their height for improving the global behavior of the building, *etc....*). These aspects can be better assessed and/or integrated by means of information collected by historical investigations and the knowledge of the construction techniques typical of that area. Indeed, thanks to the comparison with similar buildings, the amount of the specific investigations to be realized *in situ* can be reduced.

The assessment of the mechanical properties of the various constituent materials (*i.e.*, masonry blocks, mortar) and of their assembling has to be investigated by means of detailed *in situ* testing [3]. The original mechanical properties of the construction materials often have been modified by the age of the building and by loading and environmental effects over time. Thus, the detection of crack patterns and/or of decay conditions of the materials is a first check for individuating potential troubles in structural elements. The mechanical properties of masonry blocks and mortar joints are usually assessed by means of destructive tests *in situ*, even if in the last years, a significant effort has been spent to use non-destructive testing techniques [4,5,6,7,8] in order to limit destructive tests to only a few points.

Investigation techniques based on *in situ* dynamic tests carried out through ambient or forced vibrations [9,10,11,12,13,14,15] represent an innovative and useful instrument to observe different aspects of historical buildings thanks to the study of their dynamic response in terms of frequencies and modal shapes [2].

Firstly, based on the comparisons of experimental evidence (mainly consisting of the principal modal shapes and the related frequencies) and predictions given by linear dynamic models, the values of elastic constants and unit weight of masonry can be easily assessed, taking into account the overall assembly of stones/bricks and mortar.

Moreover, dynamic *in situ* tests can be aimed at studying the response of the structure under seismic actions and at investigating the interaction of the examined building with adjacent constructions or with the foundation soil. This latter topic in particular is rarely accounted for in common structural modeling because in most cases, type and dimensions of the foundations or properties of the subsoil are unknown. When the foundation soil is soft, the hypothesis of a fixed-base condition for the superstructure is no longer valid and the effect of Soil-Structure-Interaction (SSI) should be accounted for. If dynamic *in situ* monitoring is carried out, the role of SSI should be properly accounted for, because measures acquired in different parts of a building may contain the effect of coupling between soil and structure.

Finally, a periodical control of the main dynamic parameters of the building by *in situ* dynamic tests is useful to verify if some changes in the building have occurred, especially in the case of historical ones [2].

The information collected in these preliminary steps are functional to the structural model; it is clear that, if the amount of information is huge, the modeling can be more refined and the results more reliable. For example, models by Finite Elements (FE) allow understanding the structural behavior for a fixed action, identifying the original causes of the current damage patterns, individuating the resisting mechanisms, and predicting the behavior for exceptional actions (*i.e.*, seismic action).

After the building has been modeled, linear analyses under gravity loads are useful instruments to verify the accuracy of the geometrical layout and the reliability of the values of unit weight by comparing the theoretical stress distributions with the experimental data obtained *in situ* in some significant points (*i.e.*, by means of flat jacks). Moreover, a 3D model allows identifying local problems related to stress concentration by comparing them with actual situations and taking into account a rehabilitation design. As previously cited, FE modeling techniques are also a useful support to develop linear dynamic analyses and to assess the mechanical parameters of materials by means of comparison with *in situ* dynamic tests.

Modern code provisions advise to assess the safety of masonry buildings taking into account the non-linear behavior for masonry, by means of 'push-over analysis' carried out thanks to more or less refined models (methods based on storey mechanisms, equivalent frame approaches, two- or three-dimensional continuum approaches [16,17,18,19,20,21,22]. Moreover, pushover analyses are frequently used to assess the seismic vulnerability of historical masonry constructions [23] because the effort spent for the detailed modeling is compensated by the relevance of the structure.

However, the post-elastic behavior of masonry materials is quite uncertain to define, but it is a key aspect to simulate the non-linear behavior of the structure and, thus, needs to be assessed especially for seismic verifications. However, because affordable non-linear FE analyses applied to a continuum require high expertise and should be used when the knowledge of the structure is very detailed, several methods based on simplified element meshing have been developed. The reliability of these approaches for the prediction of the actual behavior of each structure strictly depends on its geometrical configuration. Moreover, in structures endowed of particular architectures (churches, arches, domes, *etc.*....) it is often possible to identify sub-structures characterized by well-defined resistant mechanisms, so that single parts of the building are verified separately [22,24,25].

In this paper a complete approach for the study of an existing masonry building with the final aim to evaluate its safety conditions is followed by analyzing a typical case study: a historical building (Palazzo Bosco Lucarelli) located in Benevento (Italy), a zone of high seismic hazard in the core of the Apennine.

THE CASE STUDY OF A HISTORICAL BUILDING

History and Description of the Structure

For the examined building, 'Palazzo Bosco Lucarelli', an old noble palace located in Benevento (Italy), research in the archives of City Hall and in public libraries allowed collecting information about the different construction phases of the building thanks to historical reports, maps, urban plans and, for the XX century, photos.

The original nucleus of the current building probably goes back to the XI century; successively, some documents of XIII and XIV century confirm the presence of a construction with an adjacent church in the same site of the current palace. This building was severely damaged during the strong earthquake of 1702, after that it was rebuilt and amplified assuming the present 'palace' structure with an internal closed court. Some maps of Benevento of XVIII century (Figure 1a) show a central hallway, the entrance on the current back prospect of the palace (South-West prospect), and the adjacent church. In the years 1768–1774 the building was retrofitted and part of the church was incorporated in the construction, such as indicated in some maps of the beginning of XIX century and in the urban plan of the town dating back to 1880 (Figure 1b). At the end of XIX century, another floor was added and in 1926, when the church was completely demolished, the west wing of the building was doubled and the current geometrical configuration was assumed (Figure 1c, Figure 2a). Nowadays the court is not central anymore, but is located asymmetrically with respect to the west wing that is double the size compared to the east one (see the plan of the ground floor in Figure 2b).

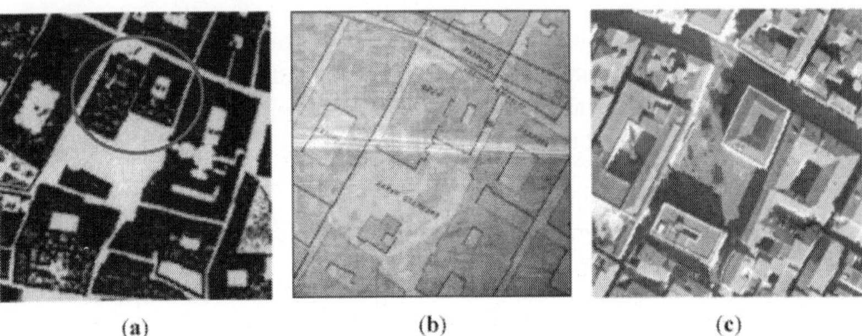

(a)   (b)   (c)

Figure 1. (a) Map of Benevento at the end of XVIII century; (b) regulator plan of Benevento in 1880; (c) aerial view of historical center of Benevento near Palazzo Bosco Lucarelli in 2012 from a satellite [26].

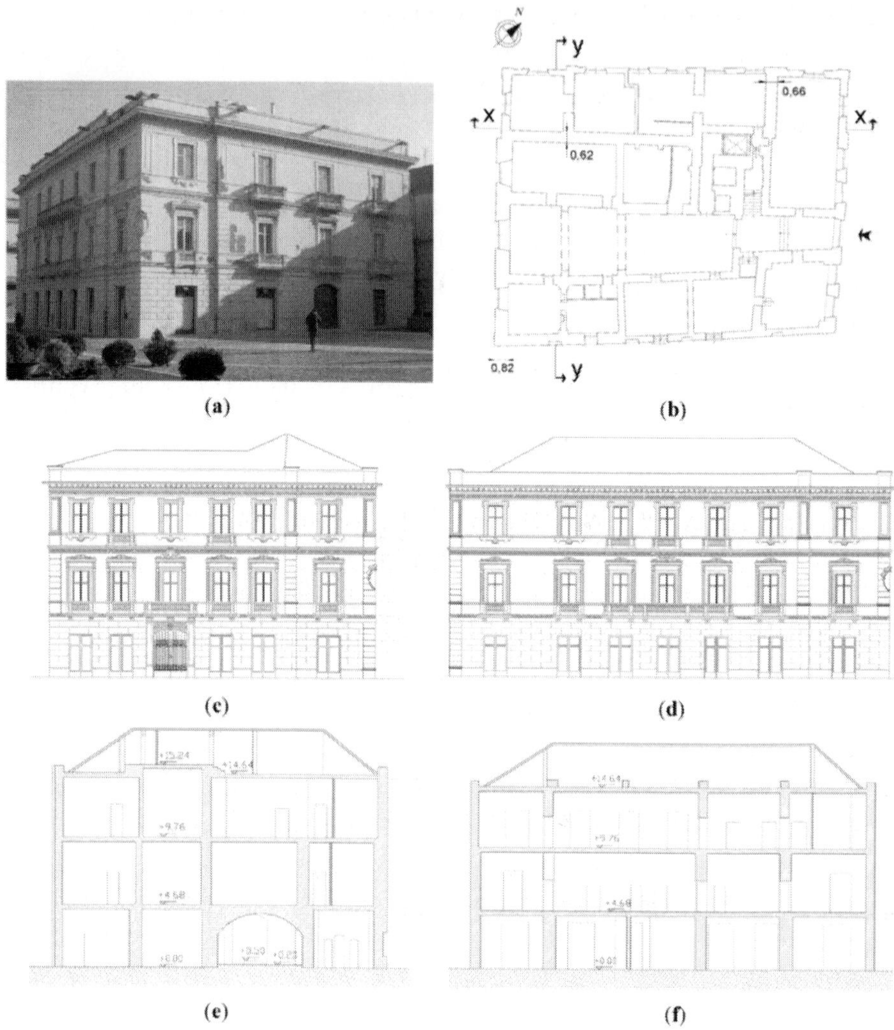

Figure 2. Palazzo Bosco Lucarelli: (a) Current photo of the Northwest and Southwest wings of the building; (b) plan of the ground floor; (c) survey of the Northeast façade (main entrance); (d) survey of the Northwest façade; (e) longitudinal section Y-Y; (f) longitudinal section X-X.

The last interventions were carried out at the end of the 90s when the building became the headquarters of the Engineering Faculty of the University of Sannio. No substantial modifications were carried out in the structural or external architectural layout, only a functional redistribution of the rooms was realized.

Nowadays the structure has a rectangular holed floor plan (Figure 2b), consisting of an underground floor, which is not completely accessible, a ground floor, two upper levels, and an attic under the pitched roof. The largest dimensions in the plan are 20 m and 16 m, the total height is 18.2 m.

The thickness of the walls varies in the range of 0.3 m–1.0 m with the greatest values attained at the ground floor (Figure 2b). A complete geometrical and structural stwo main façades (northeast and northwest prospects, Figure 2c,d) and two orthogonal sections (Figure 2e,f) are reported to illustrate the geometry of the building.

(a)   (b)   (c)

Figure 3. (a) Mixed limestone and tuff masonry at the ground floor level; (b) tuff panel at the ground floor level; (c) Clay bricks at first and second floor levels.

The original floors in most cases have been substituted by double standard 'T' shaped steel profiles with a height of 200 mm (standard type IPE 200) and spaced at a distance of 0.85 m interspersed with bricks and covered by a concrete deck 100 mm thick. Only the ceilings of the staircase and of some rooms in the entrance hall on the ground floor are still made of the original masonry vaults. The floors can be considered as rigid diaphragms and some local tests evidenced a good connection with the vertical masonry walls (reinforced concrete joists are present along the upper perimeter), so that out-of-plane mechanisms can be neglected. The connections were probably improved when the old original floors were substituted by the current steel-concrete ones.

Moreover, some steel chains are present below the first floor between the outside walls and the internal walls of the closed court; these chains were positioned at the beginning of the XX century when the last significant structural and geometrical changes of the building were performed (doubling of the west wing and construction of a new level) in order to improve the global behavior of the masonry walls between the ground and the first floor. Nowadays no sign of out-of-plane mechanism can be observed in the structure.

The roof is made of a steel truss that is supported on a perimetric reinforced concrete joist. A profiled steel sheeting, externally covered with brick tiles, closes the structure.

Assessment of Masonry Properties

The *in situ* surveys allowed detecting a masonry structure made of different materials and textures at the three levels, due to the several construction phases of the building.

At the underground and ground levels, the walls are made of irregular blocks of limestone and conglomerates (Figure 3a) with some inclusions of tuff (Figure 3b) or bricks, probably used in the past to close existing openings. At the underground level, blocks are greater and more irregular suggesting that the underlying foundations were realized as further enlargement of the masonry walls. The first two levels probably date back the construction age of the original nucleus in the IX century, since the masonry texture is typical of coeval palaces present in Benevento and dating to the same period.

At the higher levels, the walls are made of clay bricks (Figure 3c) and were probably built at the end of XIX century, as confirmed by similar masonry textures evidenced in other coeval fabrics and by the information collected about the history of the building.

At each level the walls are covered by a reinforced cement plaster (with a thickness of 50 mm and with a grid of steel bars with a diameter of 4 mm and spaced at a distance of 150 mm in two orthogonal directions, see Figure 3b,c) which can be dated back to the XX century.

Few coring examples of masonry were extracted from the walls and endoscope tests were carried out in the corresponding drilling holes; the images evidence that the walls are made of blocks along the entire thickness, without spacing between the external panels.

Compressive tests were carried out on some single blocks extracted from the masonry. The compressive test on a calcareous stone sample (cube with a side length of 50 mm) extracted at the ground floor level provided a strength of 11.6 MPa, while tests on cubes made of bricks (side length of 60 mm) extracted from the walls at the first and second levels provided strengths of 3.7 MPa and 2.7 MPa (mean values of two tests), respectively. Penetrometer tests on the mortar of the joints gave a compressive strength of about 0.65 MPa for the ground floor and 0.40 MPa for the superior levels; these values allow to classify the mortar as being of a poor typology [27,28].

However, because these data refer to single stones and, thus, are not sufficient for a complete mechanical characterization of the whole masonry assembling, the main mechanical properties (compressive strength, f_m, shear strength without vertical loads, $\tau_{o,m}$, and tensile strength, f_{tm}) were assessed based on the indications given by some recent national guidelines [28] (Table 11.D1, Appendix 11.D) because they provide lower values compared with the ones indicated in the applicative document [29] joined to the current Italian code [27]. The mechanical properties are provided as function of the masonry typology identified *in situ* by the surveys of materials (Table 1). In particular, the masonry of the ground and underground floors has been identified in the category *"Cut stone masonry with good bonding"*, while for the higher levels of the building, the masonry has been identified as being made of *"clay bricks with lime mortar"*. Considering an intermediate level of knowledge [27,28,30], achieved thanks to the *in situ* surveys, the central value of the interval reported in [28] was adopted for the compressive and shear strength. Then, an amplifying factor of 1.5 due to the beneficial effect of the reinforced plaster [27,29] and a reduction factor of 1.2 (confidence factor for taking into account the knowledge level) were considered to calculate the final values reported in Table 1.

Table 1. Mechanical properties of masonry estimated for 'Palazzo Bosco'

Floor	Masonry Typology	f_m [MPa]	$\tau_{o,m}$ [MPa]	$f_{tm}=1.5\,\tau_{o,m}$ [MPa]	W [kN/m^3]	E [MPa]	E·1.5 [MPa]
Ground	*Cut stone masonry with good bonding*	2.19	0.081	0.122	20	1500–1980 (1740)	2250–2970 (2610)
1st–2nd	*Clay bricks with lime mortar*	2.88	0.095	0.143	24	1200–1800 (1500)	1800–2700 (2250)

For the estimation of the unit weight, the samples extracted from masonry provided values of 26 kN/m^3, 17 kN/m^3 and 19 kN/m^3 for the calcareous stone, the bricks and the mortar, respectively; moreover, for the reinforced plaster a unit weight of about 24 kN/m^3 can be assumed. Based on these values and on the observed brick texture, for the masonry of the first and second floors the unit weight has been estimated to be about 17.5 kN/m^3, which is similar to the value of 18 kN/m^3 suggested in both the Italian guidelines [28] and code [29]. If the layer of reinforced plaster is considered too, the unit weight increases up to about 20 kN/m^3.

Similarly, for the irregular masonry of the ground and underground floors, the unit weight was estimated to be 22 kN/m^3, which is comparable with the value of 21 kN/m^3 suggested in both the Italian guidelines [28] code [29]. If the layer of reinforced plaster is considered too, the unit weight increases up to about 24 kN/m^3.

On the contrary, it was not possible to develop an experimental evaluation of the Young's modulus, E, of the masonry. Literature studies [31] show that the Young's modulus of masonry assembling is affected by a large variability even within the same type of stone (values of CoV also greater than 40%). Some attempts have been made to furnish empirical correlations between the Young's modulus and the compressive strength, but these correlations are a function of the stone type [32,33]. Both national guidelines and code [28,29] were considered for the assessment of Young's modulus. The same range of values (1500–1980 MPa) is given for the *"Cut stone masonry with good bonding"* of the ground floor. Conversely, for the *"clay bricks with lime mortar"*, [28] and [29] suggest the ranges of 1800–2400 MPa and 1200–

1800 MPa, respectively. The lower range suggested by [29] was considered for the *"clay bricks with lime mortar"*. Moreover, both codes specify that in case of historical masonry, the suggested values refer to masonry with poor characteristics, thick mortar joints, lack of regular horizontal layers and tidy texture, lack of transversal connecting elements between the internal and external panels of the same wall. Thus, the provided values should be reduced or increased, depending on the current conditions of the masonry. In particular, for the examined masonry types, the presence of the reinforced plaster hence also allows the use of an amplifying factor of 1.5 for the elastic modulus, as is done for the strength. In Table 1 the ranges of E for both types of masonry are reported together with the corresponding central values, assuming an intermediate level of knowledge (2610 MPa and 2250 MPa for the masonry of the ground and the upper floors, respectively). Note that these values represent the starting points for the numerical linear dynamic analyses, aimed at back-calculating the values of E by comparison between experimental and predicted values of frequencies. The above-mentioned values refer to un-cracked condition for the masonry and, thus, should be reduced considering the effective building loading history.

Characterization of the Foundation Soils

To characterize the foundation soil, a few *in situ* tests carried out near the building were interpreted. In particular, the following tests were considered: (i) a borehole carried out by the local administration very close to the building (SG-003, Figure 4a); (ii) one standard penetration test (SPT) and (iii) one down-hole test (DH). The borehole SG-003 is very deep and reaches a maximum depth of 70 m below ground level (Figure 4b). After the first six to seven meters of man-made ground and historical ruins, the subsoil is made of almost nine meters of fine-grained materials (silt and clay) followed by an alternation of coarser soils, *i.e.*, conglomerate and sandstone up to 60 meters below ground level. The succession ends with grey-azure clay (Pliocenic Age), which is typical of the Benevento area [34] and represents the bedrock formation from the viewpoint of seismic wave propagation. With the velocity of the S-waves in the first 30 meters of the subsoil, V_{S30}, was estimated around 700 m/s. This value allows classifying the site where Palazzo Bosco is located into the category B [27,30].

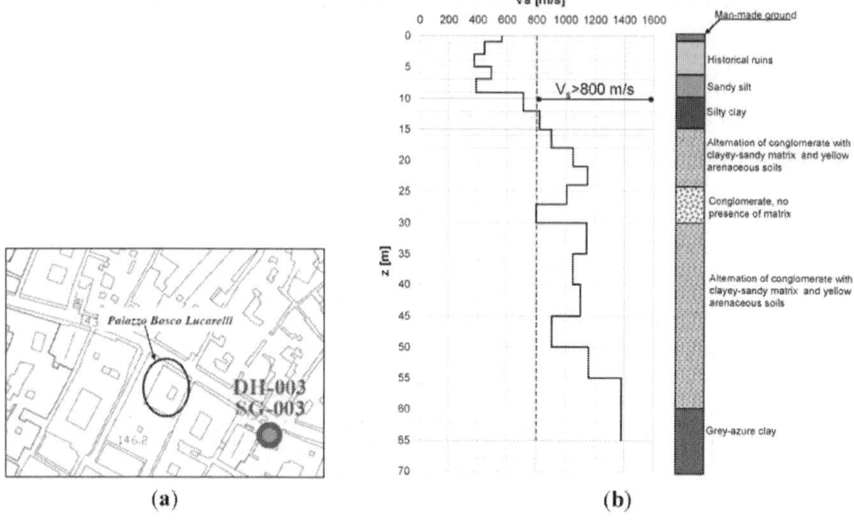

Figure 4. (**a**) Location of the boreholes and down-hole (DH) test; (**b**) shear wave velocity *vs.* depth measured by DH test.

In Situ Dynamic Monitoring

A continuous monitoring system made of nine couples of mono-axial accelerometers was placed on the three floors of the building and a three-axial accelerometer was located on the ground floor. Each couple of sensors is able to measure separately along the X and Y direction. The position of the instruments is reported in Figure 5. The sensors were aligned as accurately as possible along three vertical lines: one corresponding approximately to the center of the building and the other two to the external edges (seeFigure 5). On the first floor, the sensors were placed at about 100 mm from the first floor intrados, on the second floor at about 100 mm from the second floor extrados, and on the third floor at about 100 mm from the third floor extrados.

Figure 5. Position of accelerometers in the building.

Different sets of accelerometric recordings are available caused by ambient noise and impulsive sources (*i.e.*, the fall of 1 m^3 concrete block on a truck placed close to the building, the rapid braking of the truck along the Y direction of the building, some knocks of an instrumented hammer).

In particular, the fall of the concrete block originated vibrations in the building with well-defined peaks that were used to have a first rough estimation of the building's dynamic response. The frequency domain representation of such recordings (acquired with a sampling frequency of 500 Hz) was carried out by performing the Fast Fourier Transform (FFT) during a signal-window of eight seconds, centered on the peak value. The energy content of the recorded signals was low at frequencies higher than 50 Hz; for this reason, a low-pass FIR (Finite Impulse Response) filter was adopted in correspondence of 50 Hz. A standard baseline correction was also carried out on the original accelerometric data.

Table 2. Values of frequencies identified by impulsive measurements.

Sensors	Eigenfrequencies [Hz]	
	X	Y
A1	4.8	4.2
A2	4.8	4.2
A3	4.5	4.3
A4	4.8	4.2
A5	4.8	4.2
A6	4.5	4.3
A7	4.8	4.2
A8	4.7	4.2
A9	4.5	4.4
Mean	4.69	4.24
St. Dev.	0.15	0.07
CoV	3.1%	1.7%

The values of the first two frequencies of the building (in X and Y direction) identified by each couple of sensors are listed in Table 2 together with the mean value, the standard deviation and the Coefficient of Variation (CoV) of the measured data for each direction.

The frequencies are low scattered among the measures of the nine sensors in both X and Y direction with very small CoVs (3.1% and 1.7% for the X and Y direction, respectively). The frequency identified for the X direction is slightly greater (4.69 Hz *vs.* 4.24 Hz, +10%) than the one associated with the Y direction.

3D FE MODEL AND NUMERICAL ANALYSES

Linear dynamic analyses can be a useful instrument for identifying the first vibration modes of a building in order to better evaluate the distribution of equivalent horizontal seismic forces in case of the design of a new structure or the assessment of existing ones based on linear static analyses.

Moreover, for historical buildings, the choice of a reliable behavior factor [27,30] for linear static analysis can be very difficult and aleatory, so that non-linear static analysis can prove useful in numerically estimating the actual behavior factor.

In the following sections, both linear dynamic analyses, aimed at assessing the dynamic behavior of the building, and non-linear static analysis (pushover analysis), aimed at estimating the capacity displacement of the building and its behavior factor, will be illustrated.

Model Description of and Definition of the Loading Conditions

Based on the geometrical survey of the building, a three-dimensional Finite Element Model was implemented with the software DIANA TNO (release 9.4 [35]). The model is made of three-dimensional 'brick' elements, about square shaped in the X-Y plane. The side dimension in the X-Y plane is about 0.3 m which corresponds to half, the third or the whole thickness of each masonry wall, with the masonry thickness variable being in the range of 0.3–1.0 m. The third dimension along the Z-axis is at maximum equal to 1.0 m and allows having a maximum shape ratio between the sides of about three.

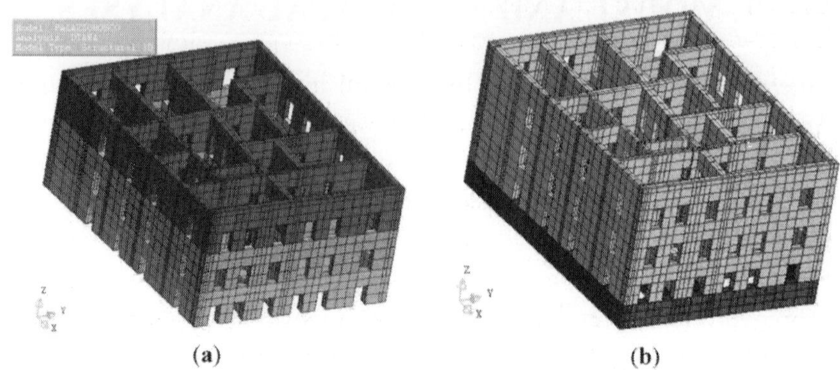

Figure 6. 3D Finite Element (FE) Model of Palazzo Bosco Lucarelli: (**a**) without the underground floor (Model A); (**b**) with the underground floor (Model B).

The 3D model is made of about 20'000 elements (Figure 6); the three floors were modeled by means of a stiff constraint in each plane and the vertical loads acting on the floors were applied directly on the corresponding masonry walls. The building was modeled both with (Model B, Figure 6b) and without (Model A, Figure 6a) the underground floor according to the type of analysis, as will be explained in the following.

As gravity loads the self-weight of masonry walls was computed using the unit weight reported inTable 1; the dead load of the floors was estimated as $G_k = 5.40$ kN/m² , while for the accidental loads, at the ground level, the first and the second floor, the values $Q_k = 3.00$ kN/m² for eventually crowded ambient [27] were considered and at the level under the roof the lower value of $Q_k = 0.50$ kN/m² [27] was assumed. The weight of the steel roof was estimated to be 1.75 kN/m², while the weight of the masonry stairs was neglected. The loads of the balconies were applied directly on the external walls ($G_k = 3.60$ kN/m², $Q_k = 4.00$ kN/m²).

Assuming a reduction factor for the accidental loads of $\psi_2 = 0.6$ under the quasi-permanent loading condition [27,36], the total weight is about 68700 kN for Model A and 88900 kN for Model B.

Linear Dynamic Analysis

Linear dynamic analyses were developed through the 3D FE Model considering all the vibration modes characterized by a participating mass ratio greater than 5%. The aims of such an analysis and its comparison with the dynamic *in situ* tests are multiple, even if they are correlated with each other: (1) to identify the dynamic behavior of the building in terms of main vibration shapes and frequencies; (2) to confirm and/or assess the reliability of the unit weight and the Young's modulus of masonry previously estimated; (3) to have indications about the interaction with the foundation soil. These topics are strictly correlated because the values of the main frequencies depend both on the mechanical properties of masonry (unit weight and Young's modulus) and on the soil-structure interaction effects. Assuming that the geometrical survey of the building is very detailed and conforms to reality and that the estimation of the unit weight of masonry is reliable (this means the global mass of the building is well-defined), the dynamic analysis will be used to assess only the Young's modulus and estimate the effect of the soil-structure interaction, if present. This latter point has been investigated thanks to some information available about the mechanical properties of the foundation soil (see Section 3.3). In order to achieve these results, (1) the mean values of the first two frequencies estimated by the dynamic measurements *in situ* (see Table 2) were used as target for the numerical values of frequencies given by the dynamic analyses; (2) the Young's modulus of masonry were varied in reliable ranges.

Firstly, the effect of the Young's modulus was investigated and, in order to take into account the cracking phenomena as suggested by the codes, the central values of the ranges reported in Table 1(Young's modulus for the ground floor $E_G = 2610$ MPa, Young's modulus for the 1st and 2nd floor and $E_{1-2} = 2250$ MPa) were reduced. For reinforced concrete structures it is in general suggested to reduce the inertia of the element by 50%, taking into account the cracking phenomena [27]; this hypothesis can be simply introduced into the modeling by reducing the Young's modulus of concrete. For masonry structures it is expected that the reduction of the inertia due to cracking phenomena is lower because of the high compressive stresses usually acting on the masonry walls. Thus, three reductions, lower than 50%, of the Young's modulus were assumed in order to: (1) verify the variation with E of the theoretical values of frequencies given by the FE model, and (2) compare them with the experimental ones:

Case 1: - 30% (E_G = 1850 MPa and E_{1-2} = 1550 MPa);
Case 2: - 45% (E_G = 1450 MPa and E_{1-2} = 1250 MPa);
Case 3: - 15% (E_G = 2200 MPa and E_{1-2} = 1850 MPa).

For these analyses in both Models A and B, the basement was considered completely restrained. Moreover, for the model with the underground level (Model B, Figure 7b), the pressure of the lateral soil was considered as acting on the underground walls by means of an earth pressure coefficient k_o at rest equal to 0.7 (no kinematism was hypothesized).

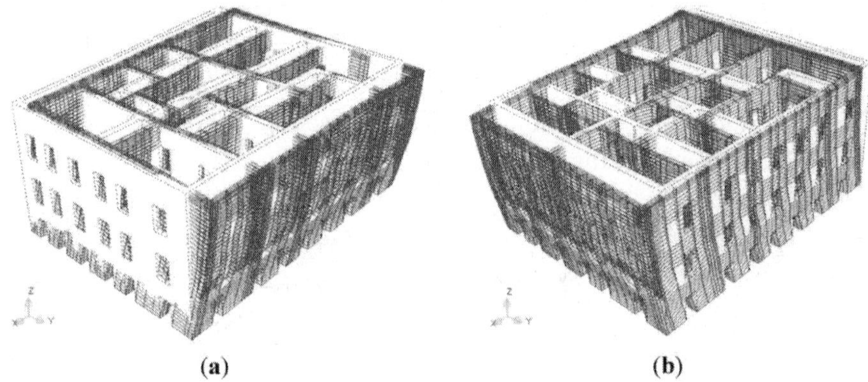

(a)   (b)

Figure 7. Modal shapes given by the 3D model for the building without the underground floor (Model A): (**a**) Mode I (Y direction); (**b**) Mode II (X direction).

The numerical values of the first four frequencies are listed in Table 3 for the three cases examined, while the shapes of the first two vibration modes are reported in Figure 7 and Figure 8 for Models A and B of the only Case 1, since they are identical for Cases 2 and 3.

Table 3. Experimental and theoretical main frequencies of Palazzo Bosco for different values of Young's modulus of masonry, E, in the case of fixed basement.

	$f_{th,fix}$ [Hz]						f_{exp}
	Case 1		Case 2		Case 3		
	E_G = 1850 MPa $E_{1\text{-}2}$ = 1550 MPa		E_G = 1450 MPa $E_{1\text{-}2}$ = 1250 MPa		E_G = 2200 MPa $E_{1\text{-}2}$ = 1850 MPa		[Hz]
Mode	Model A	Model B	Model A	Model B	Model A	Model B	Exp.
1 (Transl. Y)	4.88	4.21	4.36	3.75	5.33	4.60	4.24
2 (Transl. X)	5.45	4.71	4.87	4.19	5.96	5.14	4.69
3 (Local)	11.61	10.81	10.38	9.64	12.71	11.81	-
4 (Local)	13.08	11.58	11.67	10.32	14.29	12.64	-

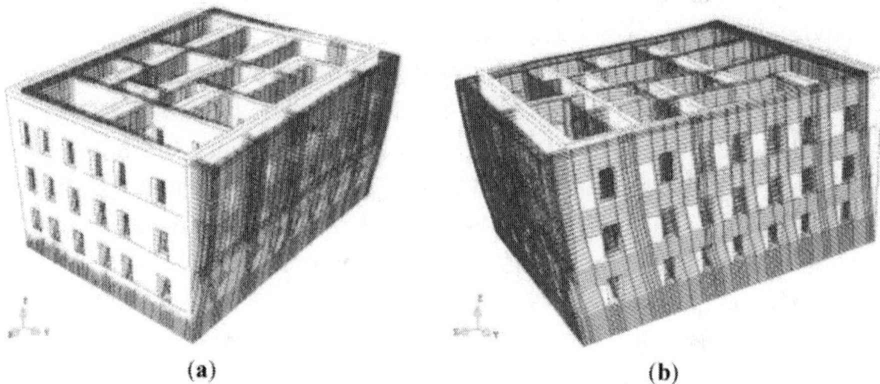

(a)   (b)

Figure 8. Modal shapes given by the 3D model for the building with the underground floor (Model B): (a) Mode I (Y direction); (b) Mode II (X direction).

In both models, the first and the second modal shapes are completely translational in direction Y (Figure 7a and Figure 8a) and X (Figure 7b and Figure 8b), respectively, with a participating mass ratio of about 75% for each direction. In agreement with the experimental results, the first frequency (direction Y) is always about 10% lower than the second one (direction X); this evidences a tendency of the building to be more deformable in direction Y that is parallel to the shorter side (East-West direction, see Figure 2b).

In the model with the underground level (Model B), the first two frequencies are reduced by about 16% with respect to Model A, clearly indicating a greater deformability of the building when the underground floor is modeled too.

The third modal shape, having a share of participant mass equal to 10 % only, is local and regards a wall of the Northwest façade (Figure 2d) on the second floor. This wall is particularly deformable due to the presence of many openings in comparison with the other façades (the two parallel façades along the X direction are not perfectly symmetrical, see plan of the building in Figure 2b). The fourth mode is local too and characterized by mass participant ratios lower than 5%.

When the effect of the Young's modulus is reduced by about 20% (*i.e.*, Case 2 *vs.* Case 1) both first and second frequencies reduce by 10%. Adopting Young's moduli lower than those assumed in Case 2 is unrealistic, because very small numerical frequencies are obtained, and, thus, they are not reliable with the experimental evidence. Moreover, any interaction with the foundation soil has not been considered yet in these analyses, and it is known that the SSI can induce a further reduction of the main frequencies. Similarly, when the Young's modulus increases by about 20% (*i.e.*, Case 3 *vs.* Case 1), the theoretical frequencies increase by about 10%. In conclusion, as expected, a relevant sensitivity of the frequencies to the variation in Young's modulus has been observed: It can be concluded that, for the examined building, a variation of 50% in E leads to a variation of about 25% in terms of frequencies.

In the last column of Table 3, the mean values of the first two frequencies experimentally indentified are listed too. As confirmed by the numerical FE analyses, higher frequencies have not been experimentally identified, probably due to the low rate of participating mass corresponding to these modes and the low intensity energy of the dynamic source. The experimental frequency of the first mode is the mean value of the values corresponding to the first peak of the FFT for each sensor measuring in the Y direction; the same applies analogously for the X direction.

Considering Model A (without the underground level), the results of Case 2 are closer to the experimental evidence (+2.8% and +3.8%). By contrast, considering Model B (with the underground level) that furnishes lower frequencies than Model A, the experimental results are best fitted by the Case 1 (−0.7% and +0.4%).

It is clear that a correct assessment of the Young's modulus of masonry should not ignore the effect of the foundation soil deformability that, in these analyses, has been completely neglected in both Models A and B because the structure has been considered fixed at the basement.

Effect of Soil Stiffness

In Section 3.2 it has been observed that for Model B, where the underground floor is modeled and lateral pressures of the subsoil are applied on the perimetric underground walls, the first two frequencies decrease by about 16% with respect to Model A (see Table 3), independently of the Young's modulus of masonry.

A first simple approach to consider also the deformability of the foundation soil consists in substituting the fixed constraint at the basement level with vertical elastic springs identified by a coefficient of sub-grade reaction, k [37]. This approach, even if it is widely adopted by designers to take into account soil deformability in the global modeling of the structures, is strongly influenced by the values assigned to the parameter k. Note that an uncorrected estimation of k could lead to an excessive and unrealistic deformability of the structure.

For the case at hand, several values of k have been assigned to the vertical springs placed under the basement walls, and it was found that for $k = 0.10$ N/mm^3 the same results of the completely fixed-base schemes are obtained. This led to consider $k = 0.10$ N/mm^3 as an upper limit of the variability range of this parameter. Moreover, varying k in the range of 0–0.10 N/mm^3, it was found that the main frequencies reduce as k decreases with a quite linear trend: $i.e.$, for $k = 0.05$ N/mm^3, that corresponds to 50% of the upper limit, the frequencies reduce by about 25% compared to the fixed-base schemes. Notwithstanding the simplicity of this approach, the parametric study enhances, thus, that the effect of k is similar to that given by the variation of the masonry Young's modulus discussed in Section 3.2. This aspect highlights how soil deformability could be important in modeling the structure response. However, different and more refined approaches for taking into account Soil-Structure Interaction exist and their use for the case at hand has been illustrated in [38].

Analysis under Gravity Loads

By means of the 3D FE model of the building, the stress distribution under gravity loads was estimated in order to check the eventual presence of anomalous stress concentrations. The gravity loads were defined in Section 3.1 and were amplified by the coefficients $\gamma_g = 1.3$ and $\gamma_q = 1.5$ for permanent and accidental loads, respectively, in order to carry out verifications at the ultimate limit state [27,39].

In Figure 9 the distributions of the vertical stresses (σ_{zz}) in the entire building (Figure 9a) and at the ground floor (Figure 9b) are reported for Model A. The highest values of the compressive stresses are 0.85 MPa for the ground floor, 0.70 MPa for the first, and 0.45 MPa for the second one. These values, referring to the strength reported in Table 1 (2.20 MPa for the ground floor and 2.70 MPa for the superior ones), correspond to safety factors of 2.6, 3.9 and 6.0, respectively. However, these values are singular peaks relieved at corners or openings, while the stresses are on average about 0.55 MPa at the ground, 0.45 MPa at the first and 0.35 MPa at the second floor.

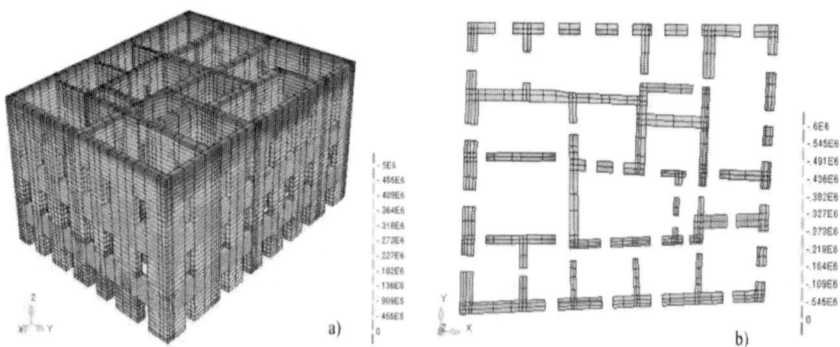

Figure 9. Vertical stress distribution, σ_{zz} (MPa, negative in compression) for Model A: (**a**) whole building; (**b**) plan of the ground floor.

When the underground level is modeled too (Model B) and the pressure of the foundation soil is taken into account on the lateral underground walls by means of the coefficient $k_o = 0.7$, a contrasting action of the soil on the transversal dilatation of the underground walls is observed. The compressive stresses are a little higher than Model A and are on average equal to 0.90 MPa at the underground level, that is, however, significantly lower that the masonry strength.

NON LINEAR STATIC ANALYSIS UNDER HORIZONTAL ACTIONS

Choice of Parameters for Non-Linear Analysis

Several non-linear static analyses of the structure under horizontal forces (Push Over) were carried out through the 3D FE model. The first purpose of these analyses was the estimation of the sensitivity of the global structural behavior to changes of some parameters typical of masonry constitutive laws, especially in terms of stiffness in the elastic field and ductility in the post-elastic field. Then, the performances of the building under horizontal actions were evaluated.

The masonry was modeled as an equivalent continuous and homogenized material (Poisson coefficient 0.3) adopting the 'total strain' model [40] with 'fixed cracking' for the tension behavior. The 'fixed crack' concept considers that the coordinate system along which the constitutive law is kept in a fixed position, defined upon cracking. Thus, the reference system to evaluate the principal stresses is defined by the first crack plane and, after cracking, a reduction of the shear modulus of the material, G, can be considered by means of a shear retention factor, β [41]. The hypothesis of $\beta = 1$ corresponds to not consider any reduction of G after cracking and, in general, leads to overestimation of the global strength of the investigated element or structure. By contrast, when the shear retention factor reduces, the strength reduces too, and the influence of the other parameters of the constitutive laws can become less significant due to the higher decay of G. For the examined building, the effect of three values of β was investigated: 1, 0.5, and 0.01.

Regarding the constitutive laws of masonry, the same Young's modulus were considered in the linear field both for compression and tension behaviors.

For compression, a linear behavior is assumed until reaching 50% of the strength, and then the stiffness reduces according to a parabolic law up to the strength. The values of the compressive strength, f_m, are the ones estimated in Table 1 for the two types of masonry. After the peak, a linear softening is considered until a residual stress of 0.8 MPa (that is 30–35% of f_m [33]) and a maximum strain of 15‰ (see Figure 10a).

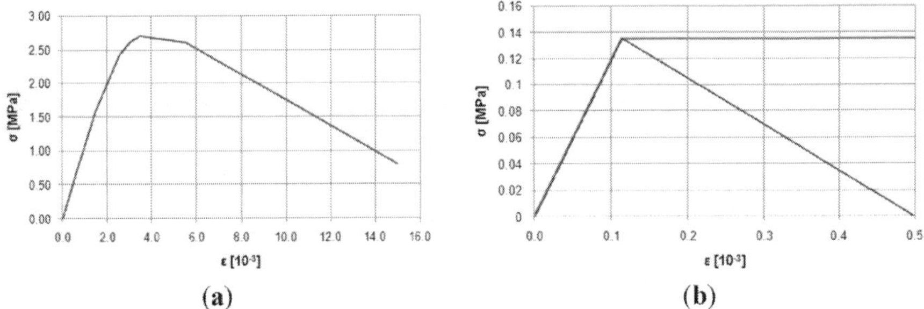

Figure 10. Constitutive relationships of masonry for Floors 1–2: (**a**) compression; (**b**) tension.

For tension, a linear behavior was assumed until the values of strength, f_{tm}, listed in Table 1 and corresponding to about the 5–6% of the compressive one, f_m, [33] were reached. After the peak, two trends were considered in order to analyze the influence of the post-elastic behavior under tension on the global ductility of the building: A linear softening of the stress until a maximum strain of 0.05% or a plastic behavior characterized by an indefinite strain with constant stress (see Figure 10b). The hypothesis of plastic behavior after the peak can be considered as an equivalent and simplified way to take into account the modeling of the homogenized material with the large slips that usually develop in the post-elastic field between masonry blocks and mortar joints.

The shape of the constitutive law in compression was kept unchanged in all the analyses, because it was expected that for a masonry building under horizontal actions the criticism was represented by the shear performances that were related to the modeling of masonry behavior in tension. In particular, it was expected that the post-peak behavior of the compressive law would not be significantly attained, so that the parametric analyses had been carried out only varying β and the post-elastic behavior under tension.

Furthermore, in order to focus the attention on this topic and remove other causes of variability in the structural behavior, the building was modeled without the underground level and under the assumption of a fixed basement (Model A). Thus, no effect of SSI was directly taken into account. However, in order to consider, even if in a simplified way, the additional deformability of the structure due to the neglected effects (underground level and SSI), the lower values of Young's modulus, $E_G = 1450$ MPa and $E_{1-2} = 1250$ MPa corresponding to Case 2, were used in the push-over analyses. Indeed, using these couple of values for E, the numerical frequencies approximate to the experimental ones (see Table 3). This choice for the Young's moduli was also determined by means of the consideration that their values influence the slope of the elastic part of the capacity curve of the structure and, thus, they must be assessed in order to replicate as well as possible the behavior of the structure in the linear field.

In Table 4 a synthesis of the main mechanical parameters assumed in the analyses for the two masonry typologies is reported.

Table 4. Mechanical parameters of masonry for non-linear analysis.

Parameters/Floors	Ground	1st–2nd
Young's Modulus	1450	1250
Compressive strength [MPa]	2.2	2.7
Tensile strength [MPa]	0.120	0.135
Ultimate strain in compression	0.015	0.015
Compressive stress at the ultimate strain [MPa]	0.80	0.80
Ultimate strain in tension—softening	0.0005	0.0005
Ultimate strain in tension—plastic	indefinite	Indefinite

Assessment of the Capacity Curve

The non-linear behavior of the structure can be synthesized by the relationship (capacity curve) between the base shear, that is the resultant of the horizontal applied forces, and the displacement of a control point of the building. The non-linear static analysis was carried out for the examined building for several reasons: (a) to estimate the displacement capacity of the building under seismic actions, (b) to assess the behavior structure factor q [27,30,36] for verifications based on linear-elastic analysis, (c) to check the sensitivity of the building response to some parameters characteristic of the masonry modeling (shear retention factor, β, constitutive law in tension). Because the first two modal shapes of the buildings are the main ones (mass participating ratio of about 75%) and are completely translational in Y and X directions with no torsional effects (see Section 3.2), the horizontal forces were applied according to a distribution proportional to the two first vibration modes of the building, separately in directions $\pm$X and $\pm$Y ('main distribution' as defined in the current Italian code [27]). The displacement of the centroid of the last floor of the building (about 15 m from the ground level) was assumed as control point for the displacement [27,36]. Note that both the assumption on the force distribution and on the control point may not reproduce correctly the actual dynamic behavior of the building; for example, an adaptive pushover procedure should furnish more reliable results [42]. The only force distributions proportional to the first two vibration modes were considered and no secondary distribution of forces, *i.e.*, an uniform pattern of loads as suggested in both national and European codes [27,30,36], since the attention was mainly focused on having a global view of the building's behavior.

In Figure 11 the capacity curves for the direction +X under the assumption of plastic behavior under tension are graphed for three values of β: 1, 0.5, and 0.01. The analyses were stopped after the base shear had attained a reduction of about 20% of the maximum value [30]. As expected, the lower β is, the lower are both the maximum horizontal load and the maximum displacement. Considering the uncertainness of the effective behavior of masonry, the lowest values for β can be reasonably assumed in order to have the safest predictions of the overall behavior of the building.

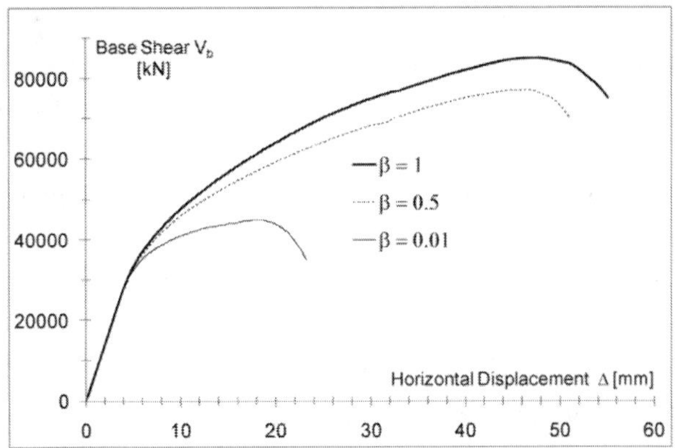

Figure 11. Capacity curves in direction +X for the three different values of β.

In Figure 12 the capacity curves in direction +X and +Y are directly compared under the hypothesis of softening or plastic behavior under tension (see Figure 10b) and assuming $\beta = 0.01$. The softening behavior does not influence the stiffness of the building in the elastic phase, but the beginning and the extension of the plastic field. Clearly, under the hypothesis of plastic behavior, the ultimate displacements increase significantly with values that are 1.6–1.7 times larger for both directions. Moreover, the strength is higher too. Because of the asymmetry of the building, the behavior in direction X and Y is not the same, but a larger deformability in the elastic phases is evidenced for direction Y, which is the shorter side and is characterized by a lower number of resistant walls as well (see Figure 2b).

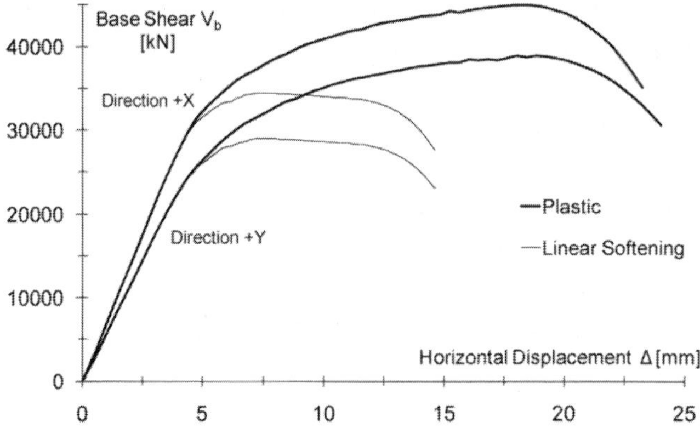

Figure 12. Capacity curves in direction +X and +Y for two different constitutive laws under tension.

In conclusion, the analyses evidences that the behavior of the whole structure is strongly influenced by the shear stiffness reduction, β, and by the softening branch of the tensile behavior. It is clear that a correct choice of β and of the post-peak behavior under tension significantly influences the safety verifications, with the displacement capacity being very different.

The exam of the strain distribution shows that, under the assumption of linear softening after the tensile strength, most of the panels are able to reach the assigned ultimate tensile strain (0.05%). Moreover, most of the compressed panels are in the elastic field because the strains are lower than 0.4%, while only in few cases the strains are proximal to the ultimate one (1.5%). These results confirm that the crisis of the structure is due to the tension behavior. By contrast, for the case of plastic behavior under tension, in most panels the tensile strains overcome the value of 0.15% that is three times larger than the maximum value (0.05%) assumed in the case of softening. This allows the entire structure to have larger displacements and to reach a slightly higher strain also under compression; nevertheless, most of the compressed panels remain in the elastic field with the strains being lower than 0.4%.

Assessment of q-Factor

On the basis of the numerical capacity curves it was possible to estimate the behavior q-factor [27,36] of the building according to the graphic procedures illustrated in Figure 13a,b for the direction +X and + Y, under the hypothesis of linear softening under tension and $\beta = 0.01$. In particular the q-factor, q_0 is defined as:

$$q_0 = q_\mu\,\rho_i \qquad\qquad \rho_i = F_u/F_l \qquad\qquad q_\mu = F_e/F_u$$

$$(2)$$

where ρ_i is the ratio of the maximum load given by the capacity curve, F_u, to the limit elastic one, F_l, and q_μ is the ratio of the theoretical elastic load corresponding to the failure condition, F_e, to F_u. The values of q_μ, ρ_i, and q_0 calculated according to the graphical procedure of Figure 13 are 2.94, 1.31 and 3.84, respectively, for the direction X and 2.89, 1.40 and 4.06 for the direction Y.

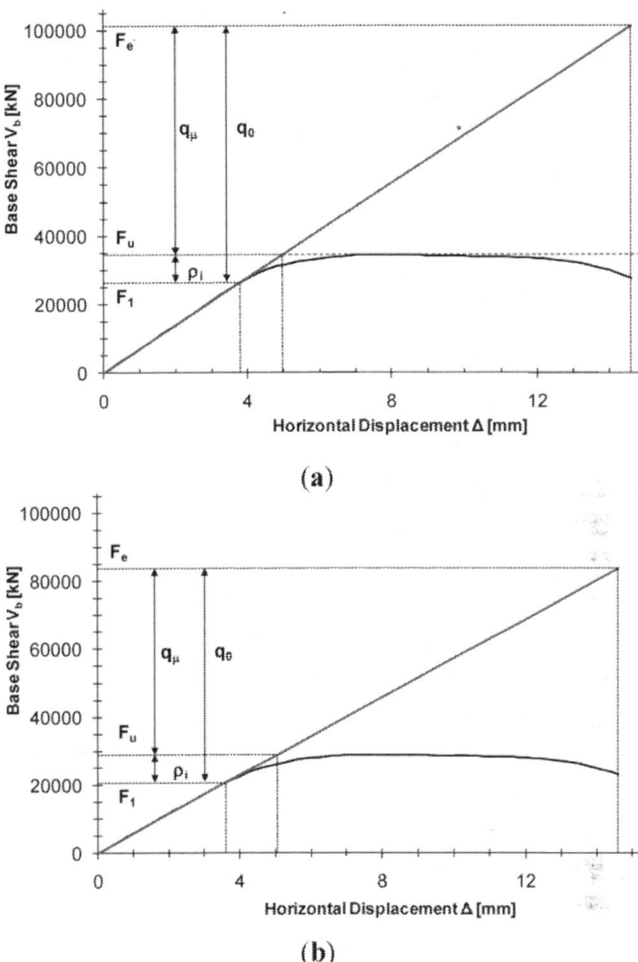

Figure 13. Graphic evaluation of q-factor in the case of linear softening after tensile strength: (**a**) direction +X; (**b**) direction +Y.

The European code [36] gives a value of q_0 ranging between 1.5 and 2.5.

The Italian code [27] gives a value of $q_0 = 2.0\ \alpha_u/\alpha_1$ for ordinary masonry constructions. Assuming $\alpha_u/\alpha_1 = 1.8$ for masonry buildings with two or more stories, as the examined one, the q-factor is $q_0 = 3.6$.

The values of q_0 calculated from the numerical curves ($q_{0x} = 3.84$ and $q_{0y} = 4.06$) are, thus, meanly 10–15% greater than the ones given by the Italian code, while they are significantly larger than the ones furnished by the European code [36]. Lower values of q_0 mean lower reduction of the seismic action in the case of linear analyses and, thus, are safer. Thus, in view of these results, the assumption of a linear softening under tension seems to be more consistent compared to the hypothesis of plastic behavior that leads to estimating a very large ductility of the structure and, thus, an even greater q-factor.

Safety Verification of the Building According to Code Indications

The capacity displacement curves reported in Figure 12 represent a characteristic of the structure and can be used to assimilate the behavior of a complex system characterized by Multi Degrees Of Freedom (MDOF system) to that of an equivalent non-linear Single Degree Of Freedom oscillator (SDOF system). Hence in the pushover procedure, after the first step leading to the definition of the capacity curve, the ultimate displacement available for the structure (MDOF system) has to be compared with a target design displacement, calculated according to national or European code indications [27,36].

The second step consists in defining the capacity displacement curve of the equivalent SDOF system that is usually assumed to have a bilinear shape. The equivalence is established by the principle that the area above the bilinear curve of the SDOF system is equal to the area above the true capacity curve of the structure. To this aim, firstly, the capacity curve of the MDOF structure has to be normalized by defining the force $F*$ and the displacement $d*$ as follows:

$$F* = \frac{F_b}{\Gamma} \qquad d* = \frac{d_n}{\Gamma} \tag{3}$$

where F_b and d_n are the base shear force and the control node displacement of the structure in its capacity curve and Γ is a transformation factor associated to the first modal shape. For the examined building the factor Γ is 1.29 and the 1st modal shape is the main one for both directions.

In Figure 14 for the direction ±X (note that the building shows a lower displacement capacity in direction ±X than in direction ±Y) the capacity displacement curves of the structure normalized to Γ and the bilinear ones of the equivalent SDOF systems are reported for both cases of linear softening and plastic behavior after the tensile strength. The minimum values of the maximum displacement $d*$ in direction $\pm X$ are equal to 11.20 mm and 17.74 mm under the assumption of linear softening or plastic behavior, respectively. This means that the maximum peak ground acceleration that the building can sustain is 0.74 g in the case of linear softening or 0.91 g in the case of plastic behavior under tension.

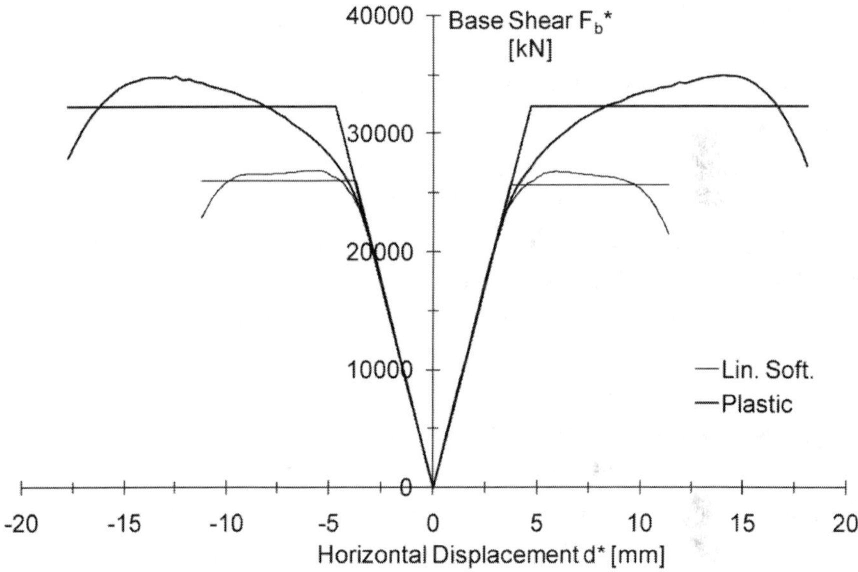

Figure 14. Capacity displacement curves of the structure normalized to Γ and of the SDOF equivalent system in direction $\pm X$ for the case of plastic behavior and linear softening under tension.

Considering that, according to the seismic risk map of Italy [43], the expected peak ground acceleration, a_g, for the site is $a_{g,475} = 0.262g$ assuming a return period of 475 years (probability of exceedance of 10% in 50 years) or $a_{g,975} = 0.325g$ assuming a return period of 975 years (probability of exceedance of 5% in 50 years), the building has a minimum safety factor of 2.8 and 2.3 for these two values of acceleration, respectively.

These values of peak ground acceleration can be modified according to an importance factor γ_I that in this case can be assumed to be equal to 1.0 considering the building in Class II (the classes are significant to the destination importance of the building and are numbered from I to IV); indeed the building has no importance for public safety in view of the consequences associated with a collapse, and its integrity is not of vital importance for civil protection during earthquakes [27,36].

CONCLUSIONS

The structural analysis of historical masonry buildings is particularly complex since each construction is a stand-alone system usually designed and erected for being a singular case. In this paper the procedure for assessing the seismic risk of a historical masonry building is applied to a case study that is of a very common typology of masonry building in Italy and, moreover, is located in an area with high seismicity. The main aim of the procedure is to identify the key parameters in the implementation of a FE numerical model aimed at allowing to carry out allowable analyses in the linear dynamic and non-linear static field.

The greatest difficulties in facing the study of the building were the definition of the mechanical properties of masonry, especially the strength and the elastic modulus. Furthermore, if the effect of the soil-structure interaction is accounted for, the type of modeling and the mechanical characterization of the soil are necessary. An important contribution to calibrating the FE linear model was given by the results of the dynamic *in situ* tests used as target of the numerical analysis in terms of fundamental frequencies. By means of the FE model, indeed, the influence of the uncertainties of masonry Young's modulus and soil deformability, which is simply taken into account by means of vertical linear elastic springs, on the dynamic behavior of the structure was studied; the following concluding remarks are highlighted:

- The influence of the deformability of the masonry and soil was found to be comparable, however, as it is well-known, the real uncertainty in evaluating soil deformability (*i.e.*, the coefficient of sub-grade reaction, k) is much higher than the one of the masonry;

- Concerting the properties of masonry, the non-linear static analysis demonstrated the major importance of the modeling of the constitutive relationship under tension, since the ductility of the entire building was directly related to the ductility developed by the masonry under tension;

- Regarding the behavior factor suggested by Italian and European codes, the numerical analysis provided higher values especially when the constitutive relationship of masonry under tension was assumed to be elastic-plastic, thus the assumption of a linear softening under tension seemed to be more adequate to obtain a safer prediction.

In conclusion, the present work enhanced the key parameters regulating the structural behavior of a historical masonry building under seismic actions (*i.e.*, the structural models) and the utility of the *in situ* dynamic tests.

Further studies are necessary to define the most suitable model to account for both the soil-structure interaction and the behavior of masonry under tension.

REFERENCES AND NOTES

1. ICOMOS Charter. Principles for the Analysis, Conservation and Structural Restoration of Architectural Heritage; International Council on Monument and Sites (ICOMOS). In Proceedings of the 14th ICOMOS General Assembly, Victoria Falls, Zimbabwe; 2003.

2. *Linee Guida per la valutazione e riduzione del rischio sismico del patrimonio culturale allineate alle nuove Norme Tecniche per le costruzioni (D.M. 14 gennaio 2008); G.U.R.I.*; 2008; Volume 24, 1 29; (in Italian).

3. Binda, L.; Saisi, A.; Tiraboschi, C. Investigation procedures for the diagnosis of historic masonries. *Constr. Build. Mater.* 2000, *14*, 199–233.

4. McCann, D.M.; Forde, M.C. Review of NDT methods in the assessment of concrete and masonry structures. *NDT E Int.* 2001, *34*, 71–84.

5. Clark, M.R.; McCann, D.M.; Forde, M.C. Application of infrared thermography to the non-destructive testing of concrete and masonry bridges. *NDT E Int.* 2003, *36*, 265–275.

6. Carpinteri, A.; Invernizzi, S.; Lacidogna, G. *In situ* damage assessment and non linear modelling of historical masonry towers. *Eng. Struct.* 2005, *27*, 387–395.

7. Binda, L.; Zanzi, L.; Lualdi, M.; Condoleo, P. The use of georadar to assess damage to a masonry Bell Tower in Cremona, Italy. *NDT E Int.* 2005, *38*, 171–179.

8. Carpinteri, A.; Invernizzi, S.; Lacidogna, G. Historical brick-masonry subjected to double flat-jack test: Acoustic emissions and scale effects on cracking density. *Constr. Build. Mater.* 2009, *23*, 2813–2820.

9. De Sortis, A.; Antonacci, E.; Vestroni, F. Dynamic identification of a masonry building using forced vibration test. *Eng. Struct.* 2005, *27*, 155–165.

10. Bennati, S.; Nardini, L.; Salvatore, W. Dynamical behaviour of a masonry medieval tower subjected to bell's action. Part I: Bell's action measurement and modelling. *J. Struct. Eng. ASCE* 2005, *131*, 1647–1655.

11. Gentile, C. Operational Modal Analysis and Assessment of Historical Structures. In Proceedings of the 1st International Operational Modal Analysis Conference, Copenhagen, Denmark, 26–27 April 2005.

12. Gentile, C.; Saisi, A. Ambient vibration testing of historic masonry towers for structural identification and damage assessment. *Constr. Build. Mater.* 2007, *21*, 1311–1321.

13. Ivorra, S.; Pallarés, F.J. A Masonry Bell-Tower Assessment by Modal Testing. In Proceedings of the 2nd International Operational Modal Analysis Conference (IOMAC), Copenhagen, Denmark, 30 April–2 May 2007.

14. Ceroni, F.; Pecce, M.; Voto, S.; Manfredi, G. Historical, architectural and structural assessment of the Bell Tower of Santa Maria del Carmine. *Int. J. Archit. Herit.* 2009, *3*, 169–194.

15. Rainieri, C.; Fabbrocino, G. Operational modal analysis for the characterization of heritage structures. *Geofizika* 2011, *28*, 109–126.

16. Tomazevic, M. The Computer Program POR; Report ZRMK, Ljubljana, Slovenian, 1978; (in Slovenian).

17. Tomazevic, M. Dynamic modelling of masonry buildings: Storey mechanism model as a simple alternative. *Earthq. Eng. Struct. Dyn.* 1987, *15*, 731–749.

18. Magenes, G.; Della Fontana, A. Simplified non-linear seismic analysis of masonry buildings. *Proc. Br. Mason. Soc.* 1998, *8*, 190–195.

19. Lourenço, P.B. Anisotropic softening model for masonry plates and shells. *J. Struct. Eng.* 2000, *126*, 1008–1016.

20. Massart, T.J.; Peerlings, R.H.J.; Geers, M.G.D. Mesoscopicmodelling of failure and damage induced anisotropy in brick masonry. *Eur. J. Mech. Solids* 2004, *23*, 719–735.

21. Calderini, C.; Lagomarsino, S. A continuum model for in-plane anisotropic inelastic behaviour of masonry. *J. Struct. Eng.* 2008, *134*, 209–220.

22. Chen, S.-Y.; Moon, F.L.; Yi, T. A macroelement for the nonlinear analysis of in-plane unreinforced masonry piers. *Eng. Struct.* 2008, *30*, 2242–2252.

23. Ceroni, F.; Pecce, M.; Manfredi, G. Modelling and seismic assessment of the bell Tower of Santa Maria del carmine: Problems and solutions. *J. Earthq. Eng.* 2010, *14*, 30–56.

24. De Luca, A.; Giordano, A.; Mele, E. A simplified procedure for assessing the seismic capacity of masonry arches. *Eng. Struct.* 2004, *26*, 1915–1929.

25. Lagomarsino, S.; Galasco, A.; Penna, A. Non Linear Macro-element Dynamic Analysis of Masonry Buildings. In Proceedings of the ECCOMAS Thematic Conference on Computational Methods in Structural Dynamics and Earthquake Engineering, Rethymno, Crete, Greece, 13–16 June 2007.

26. Aerial view of historical centre of Benevento, Italy form satellite. Available online: http://maps.google.it/maps (accessed on 15 January 2012).

27. NTC 2008. *Norme Tecniche per le Costruzioni*, Decreto Ministeriale del 14/01/2008. In *G.U.R.I.*; 2008 2 4; Volume 29, (in Italian).

28. OCPM 3431. Primi elementi in materia di criteri generali per la classificazione sismica del territorio nazionale e di normative tecniche per le costruzioni in zona sismica. *Ordinanza del Presidente del Consiglio dei Ministri del 03/05/2005. In G.U.R.I.*; 10 05 2005; 107. (in Italian).

29. Circolare 617. *Istruzioni per l'applicazione delle "Nuove norme tecniche per le costruzioni di cui al D.M. 14 gennaio 2008*. Ministero dei Lavori Pubblici, Roma, 02/02/2009 (in Italian).

30. Eurocode 8. *Design of Structures for Earthquake Resistance—Part 3: Assessment and Retrofitting of Buildings*; Committee for Standardization: Brussels, Belgium, 2005. EN 1998-1.

31. Marcari, G.; Fabbrocino, G.; Lourenço, P. Mechanical Properties of Tuff and Calcarenite Stone Masonry Panels under Compression. In Proceedings of the 8th International Masonry Conference, Dresden, Germany, 4–7 July 2010.

32. Binda, L.; Pina-Henriques, J.; Anzani, A.; Fontana, A.; Lourenço, P.B. A contribution for the understanding of load-transfer mechanisms in multi-

leaf masonry walls: Testing and modelling. *Eng. Struct.* 2006, *28*, 1132–1148.

33. Augenti, N.; Parisi, F. Constitutive models for tuff masonry under uniaxial compression. *J. Mater. Civ. Eng.* 2010, *22*, 1102–1111.

34. Improta, L.; di Giulio, G.; Iannaccone, G. Variations of local seismic response inBenevento (Southern Italy) using earthquakes and ambient noise recordings. *J. Seismol.* 2005, *9*, 191–210.

35. DIANA Finite Elements Analysis. In *User's Manual—Release 9.4—Element Library, Material Library*; TNO Building and Construction Research, Department of Computational Mechanics: A.A. Delft, Netherlands, 2009.

36. Eurocode 8. *Design of Structures for Earthquake Resistance—Part 1: General Rules, Seismic Actions and Rules for Buildings*; Committee for Standardization: Brussels, Belgium, 2005. EN 1998-1..

37. Terzaghi, K.; Peck, R.B. *Soil Mechanics in Engineering Practice*; Wiley: New York, NY, USA, 1948. [Google Scholar]

38. Ceroni, F.; Sica, S.; Pecce, M.; Garofano, A. Effect of Soil-Structure Interaction on the dynamic behavior of masonry and RC buildings. In Proceedings of the 15th World Conference on Earthquake Engineering (WCEE), Lisbon, Portugal, 24-28 September, 2012.

39. Eurocode 2. *Design of Concrete Structures—Part 1-1: General Rules and Rules for Buildings*; SPRING Singapore: Singapore, 2004. ENV 1992-1-1: 2004: E..

40. Selby, R.G.; Vecchio, F.J. Three-dimensional Constitutive Relations for Reinforced Concrete. In*Technical Report 93-02*; Department Civil Engineer, University of Toronto: Toronto, Canada, 1993.

41. Crisfield, M.A. *Non-Linear Finite Element Analysis of Solids and Structures*; John Wiley & Sons: Hoboken, NJ, USA, 1991; Volume 1.

42. Galasco, A.; Lagomarsino, S.; Penna, A. On the Use of Pushover Analysis for Existing Masonry Buildings. In Proceedings of the 1st European Conference on Earthquake Engineering and Seismology, Geneva, Switzerland, 3–8 September 2006.

43. Stime di pericolosità sismica per diverse probabilità di superamento in 50 anni: Valori di a_g. Available online: http://esse1.mi.ingv.it/d2.html (accessed on 15 January 2012, in Italian).

CITATION

Francesca Ceroni, Marisa Pecce, Stefania Sica and Angelo Garofano. Assessment of seismic vulnerability of a historical masonry building, doi:10.3390/buildings2030332.

CHAPTER 6

An Experimental Study on the Effect of Foundation Depth, Size and Shape on Subgrade Reaction of Cohessionless Soil

Wael N. Abd Elsamee

Faculty of Engineering, Sinai University, El Arish, Egypt

INTRODUCTION

Soil bearing capacity and soil modulus of subgrade reaction are some various measures of strength-deformation properties of soil. To perform the structural analysis of footings one must know the principles of evaluating the coefficient of subgrade reaction "k_s".

One of the most popular models in determining the modulus of subgrade reaction is Winkler (1867) model [1]. In this model the subgrade soil is assumed to behave like infinite number of linear elastic springs that the stiffness of the spring is named as the modulus of subgrade reaction. This modulus depends on some parameters such as soil type, size, shape, depth and type of foundation.

Iancu-B. T. and Ionut O. T. (2009) presented a numerical simulation of plate loading test in order to underlines the size effect on settlements. The obtained results are compared with Finite Element Method (FEM) using the Mohr-Coulomb soil model. The obtained numerical results revealed that the subgrade reaction coefficient was strictly dependent on the size of the loaded area and the loading magnitude [2].

Elsamny, M. K., Elsedeek, M. B. and Abd Elsamee, W. N. (2010) presented field determination of the Young's modulus "Es" of footings on cohesionless soil by using plate load test [3].

Dae. S. K. and Seong Y. P. (2011) presented plate loading tests to evaluate the compaction quality of the railroad subgrade in Korea. Two methods to determine the design modulus were used. One is an unrepetitive plate loading test (uPLT) that obtains the subgrade reaction modulus (K30) and the other is a repetitive plate loading test (rPLT) that obtains the strain modulus (Ev) [4].

Aminaton M. et al. (2012) presented Winkler model and the sub grade soil is assumed to behave like infinite number of linear elastic springs. The foundation size effect on sandy sub grade by using of finite element software (Plaxis) is presented [5].

DETERMINATION OF SUBGRADE REACTION "KS"

Determination of Subgrade Reaction "K_s" Using the Elastic Parameters "E_s, V_s"

The coefficient of subgrade reaction k_s is the ratio between the pressure "q" at any given point and the settlement "δ" produced by load application at that point.

Biot (1937), Terzaghi (1955), Vesic (1961), Meyerhof and Baike (1965), Selvadurai (1984) and Bowles (1998) have investigated the factors affect the determination of k_s. Biot (1937) solved the problem for an infinite beam with a concentrated load resting on a 3D elastic soil continuum. Biot found a correlation of the continuum elastic theory and Winkler model [6]. Vesic (1961) tried to develop a value for k_s, by matching the maximum displacement of the beam. He obtained an equation for k_s to be used in the Winkler model [7,8].

However, different formulii to calculate the modulus of subgrade reaction "k_s" by some different authors are presented in **Table 1**.

Determination of Subgrade Reaction "K$_s$" in-Situ using Plate Loading test (P.L.T)

The plate-load test provides a direct measure of compressibility and occasionally of the bearing capacity of soils which are not easily sampled. The modulus of subgrade reaction can be determined by using the plate-load test as follows:

Terzaghi's Method (1955)

A major problem is to estimate the numerical value of "k$_s$". One of the early contributions was that of Terzaghi (1955) [9]. He suggested values of k$_s$ for (1 × 1) ft rigid slab placed on a soil medium. "k$_{sf}$" for full-sized footings could be obtained from plate-load tests using the following equations:

1) For square footing on cohesionless soil with dimensions = B × B.

$$k_{sf} = k_{sp} \left[\frac{B + 0.305}{2B} \right]^2$$

(1)

2) For rectangular footing on cohesionless soil with dimensions = B × L.

$$k_{sfr} = \frac{k_{sf}\left(1 + \dfrac{B}{L}\right)}{1.50}$$

(2)

3) For long foundation [strip footing] with a width = B The modulus of subgrade reaction is approximately equal to 0.67 k$_{sf}$

where:

Table 1. Some Different formulii to calculate the modulus of subgrade reaction, Ks.

No.	Investigator	year	Suggested formula
1	Winkler	(1867)	$k_s = \dfrac{q}{\delta}$
2	Biot	(1937)	$k_s = \dfrac{0.95E_s}{B(1-v_s^2)}\left[\dfrac{B^4E_s}{(1-v_s^2)EI}\right]^{0.108}$
3	Terzaghi	(1955)	$k_{sf} = k_{sp}\left(\dfrac{B+B_1}{2B}\right)$
4	Vesic	(1961)	$k_s = \dfrac{0.65E_s}{B(1-v_s^2)}\sqrt[12]{\dfrac{E_sB^4}{EI}}$
5	Meyerhof and Baike	(1965)	$k_s = \dfrac{E_s}{B(1-v_s^2)}$
6	Selvadurai	(1984)	$k_s = \dfrac{0.65}{B}\cdot\dfrac{E_s}{(1-v_s^2)}$
7	Bowles	(1998)	$k_s = \dfrac{E_s}{B_1(1-v_s^2)m I_s I_f}$

k_{sp} = plate-load test value of modulus of subgrade reaction kN/m^3, using square plate (1 × 1) ft or circular plate with diameter = 0.305 m;k_{sf} = desired value of modulus of subgrade reaction forfull-sized square footings B × B, kN/m^3;k_{sfr}= desired value of modulus of subgrade reaction for rectangular full-sized footings B × L, kN/m^3;B = footing width, meter or least dimension ofrectangular or strip.

Different Cods

American Code (ASTM D1194) (1994) and British standards Code (BS5930) (1997):

American Code (ASTM D1194) (1994) and British standards Code (BS5930) (1997) estimated the numerical value of "k_s" from plate load test results [10].

Peck, Hanson and Thobrnburn (1997)

Peck, Hanson and Thobrnburn (1997) estimated the numerical value of "k_s" by using plate load test as fol- lows:

1) Settlement on sands occurs almost entirely during construction.

2) Maximum differential settlement between footings on sand is less than 20 mm.

3) "k_s" is calculated from the straight line portion of the load-settlement curve.

Ping-Sien Lin, Li-Wen Yang and C. Hsein Juang (1998)

Ping-Sien Lin, Li-Wen Yang and C. Hsein Juang (1998) made a series of plate-load tests to investigate the load settlement characteristics of a gravelly cobble deposit and estimate the value of modulus of subgrade reaction "k_s" as follows:

$$k_s = \frac{q_a}{\delta_a}$$

(3)

where: k_s = modulus of subgrade reaction, kN/m^3; q_a = allowable bearing capacity, kN/m^2; d_a = allowable settlement against q = q_a, meter

$$q_a = \frac{q_u}{f.s.}$$

(4)

where: q_u = Ultimate bearing capacity, kN/m^2; f.s. = Factor of safety = 3 [11].

Egyptian Code (2001)

Egyptian Code (2001) made a series of plate-load tests to investigate the load settlement characteristics and estimates the value of modulus of subgrade reaction "k_s" as follows:

$$k_s = \frac{q}{\delta}$$

(5)

where:k_s = Modulus of subgrade reaction, (kN/m^3);q = Stress at settlement =1.3 mm after ten times loaded, (kN/m^2);d = Settlement against q (meter) [12].

Reza Z. M. and Masoud J. (2008)

Reza Z. M. and Masoud J. (2008) presented a direct method to estimate the modulus of subgrade reaction by the plate load test done with 30 - 100 cm diameter circular plate or equivalent rectangular plate [13].

However, Table 2 presents some different methods using plate load test with different sizes and shapes to determinate the value of modulus of subgrade reaction, k_s.

PRESENT EXPERIMENTAL STUDY

Plate load tests have been carried out in field and the settlement of sandy soil was measured under different stress levels. In the present study each sample has been placed in an open box and compacted in layers with different relative densities. Settlement has been measured under different stresses and at different relative densities as well as different depth of foundations.

Field Samples

Graded sand (GS) at different relative densities was used in field. Each sample has been compacted in layer and the relative density for each layer has been determined by using sand cone.

Loading

The load has been applied by using steel frame fixed in the ground as shown in Figure 1. The applied load has been measured by using pressure gauge connected to a jack.

Used Plates

Nine steel rigid plates were used in the tests which are divided into three groups. The first group has rigid three circular shape plates. The second group is three rigid square plates first one having dimension (1 * 1) ft and the other two square plates having equivalent area for 455 mm and 610 diameter. The last group is a rigid three rectangular plate having the same equivalent areas as the first group. The plates have concentric marking on one face and plated against corrosion. The plates have a finished thickness of 32 mm and are according to ASTM D1194 and D1196 specification as shown in Table 3 and Figure 2.

Depth of Foundations

The settlement has been measured at surface and at different depths. The depth of foundation is considered a function of width of the plate B. Circular, square and rectangular steel boxes have been placed around the rigid steel plates to be used in case of filling soil around as surcharge for different foundation depths (0.25 B, 0.50 B, 0.75 B and 1.00 B).

Test Procedure

The test procedure is as follows:

1) The soil has been placed in a square open box.

2) The box was filled with different soil layers compacted to different densities which has been determined by sand cone test. The field compaction has been done

Table 2. Some different methods using plate load test to calculate K_s

No.	Investigator	Year	Used plate shape	Used plate dimension (size)
1	Terzaghi	(1955)	Square	[305 * 305] mm = (1 * 1) ft
2	ASTM Committee D1194 on Soil and Rock and is the direct responsibility of Subcommittee D18.08 on Special and Construction Control Tests.	(1994)	Circular	Diameter from (305 to 762 mm),
3	British standards codes (BS5930)	(1997)	Circular or square of equivalent area	Diameter from 300 mm to 1000 mm
4	Peck, Hanson, Thornburn	(1997)	Square	[305 * 305] mm
5	Ping-Sien Lin, Li-Wen Yang, and C. Hsein Juang	(1998)	Circular	Diameter from 0.75, 0.90, and 1.05 m
6	Egyptian Code	(2001)	Circular or square of equivalent area	Diameter from 0.30 - 0.45 - 0.706 m and square equivalent area 0.3 * 0.3 - 0.706 * 0.706
7	Reza Z. M. and Masoud J.	(2008)	Circular or rectangular of equivalent area	Diameter from 30 - 100 cm.

Table 3. The used plates in the experimental study.

Circular plate (diameter) B (mm)	Square plate ($B \times B$) (mm)	Rectangular plate ($B \times L$) (mm)	Equivalent area mm²	Thickness (mm)	Weight (kg)
305 mm	-	238 * 307 mm	73061.66 mm²	32 mm	14.5 kg
-	305 mm	-	93025 mm²	32 mm	18.5 kg
455 mm	403.2 mm	360 * 451.6 mm	162597.05 mm²	32 mm	33 kg
610 mm	540.6 mm	458 * 638 mm	292246.66 mm²	32 mm	56 kg

Figure 1: Loading Frame,

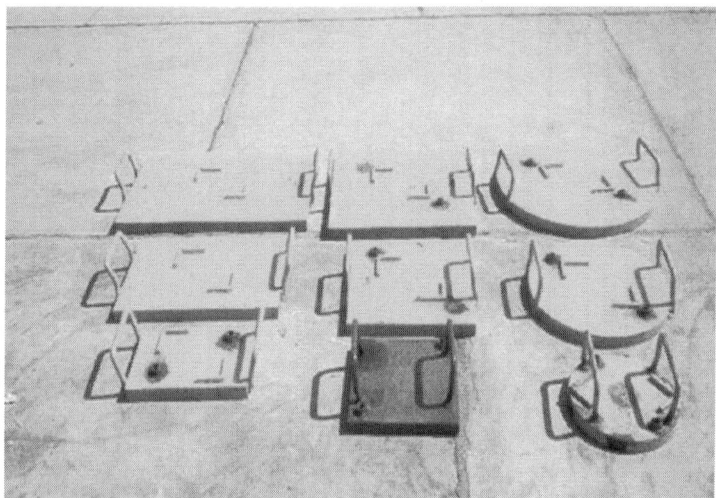

Figure 2. The Nine Rigid Plates,

Using the following:

a)   Each compacted layer has (7.5) cm thick ness.

b)   A (4.5) kg weight hammer was used and released from (30) cm height.

3) The surface of the tested soil was prepared for plate test using fine sand at the surface.

4) The steel plates were placed on the prepared surface.

5) A hydraulic jack was placed on the steel plate.

6) Four dial gauges has been placed on the plate surface.

7) The settlement has been measured by using dial gauges of sensitivity 0.01 mm placed on the edges of the steel plate. Figure 3 shows measuring settlement at surface.

8) Steel boxes have been placed around the rigid steel plates to be used in case of filling soil around as surcharge for different foundation depths (0.25) B, (0.50) B and (1.00) B as shown in Figures 4 and 5.

9) The load was applied in increments by using steel frame. Each load increment was maintained constant until the settlement rate reaches 0.02 mm/min and not less than one hour in any case.

EXPERIMENTAL RESULTS

Settlement in field was recorded for different footings

Figure 3. Settlement readings at surface using dial gauges.

Figure 4. Settlement Readings in case of surcharge using circular steel box.

Figure 5. Settlement Readings In Case Of Surcharge Using Square Steel Box.

sizes and shapes (circular, square and rectangular) under different stresses ranging between 0.589 kN/m^2 and 5.301 kN/m^2. However, the settlement has been measured at different relative densities and at different depths (0.00 B), (0.25 B), (0.50 B), (0.75 B) and (1.00 B) for all kind of plates. From the measured settlement of cohessionless soil the following relationships are obtained.

Ultimate Bearing Capacity using Experimental Results

The ultimate bearing capacity of cohessionless soil has been determined from the relationships between the stresses and the measured settlement at surface and at different depths for all plates by tangent-tangent method according Egyptian Code. Figure 6 gives an example of determination of the ultimate capacity.

Determination of Subgrade Reaction "K_s" using Experimental Results

The allowable bearing capacity (q_a) is determined by dividing the ultimate bearing capacity (q_u) by F.S. =3.0, after which the corresponding settlement (S_a) is determined. Thus, k_s is calculated based on dividing the allowable bearing capacity (q_a) by the corresponding settlement (S_a) as shown in Figure 7.

EFFECT OF FOUNDATIONS DEPTH ON SUBGRADE REACTION "KS"

The effect of foundations depth on subgrade reaction k_s has been investigated. However, the values of subgrade reaction k_s have been obtained under different plates with different foundation depths as shown in Table 4. Figures 8 and 9 show examples for the effect of footing depth on k_s for different angle of internal friction under different plate shapes. These figures show that subgrade reaction k_s of cohessionless soil increases with increasing footing depth.

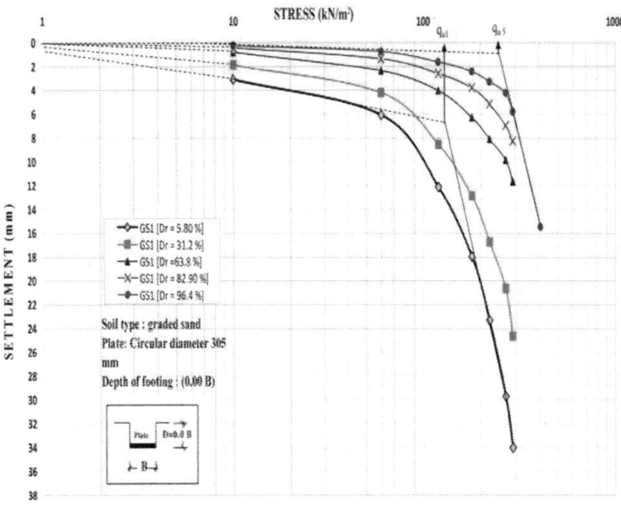

Figure 6. The Relationship Between Stress And Settlement Of Plate For Determination Of Ultimate Bearing Capacity For Circular Plate Diameter 355 Mm.

Table 4. Values of K_s (Kn/M^3) of cohessionless soil using plate load test Diameter 305 Mm, Square 305 * 305 Mm (1 * 1) Ft and Rectangular 238 * 307 Mm (Equivalent Area For 305 Mm Diameter).

Angle	Depth			
36°	0.50 B	7933.66	8489.01	9337.92
	0.75 B	8642.47	9247.44	10172.19
	1.00 B	9186.34	9829.39	10812.32
	0.00 B	6573.45	7033.60	7736.95
	0.25 B	9071.51	9706.52	10677.17
39°	0.50 B	9155.02	9795.88	10775.46
	0.75 B	10230.24	10946.36	12041.00
	1.00 B	10420.23	11149.65	12264.62
	0.00 B	8027.61	8589.54	9448.50
	0.25 B	9637.31	10311.92	11343.11
42°	0.50 B	9898.28	10591.16	11650.28
	0.75 B	11399.41	12197.37	13417.11
	1.00 B	13447.55	14388.88	15827.77

Where: B = diameter of circular plate, width of square plate or smallest dimension of rectangular plate.

Angle of internal friction (θ)	Depth of footing	k_s circular plate 305 mm diameter kN/m^3	k_s square plate 305 * 305 mm kN/m^3	k_s rectangular plate 238 * 307 mm kN/m^3
	0.00 B	1994.90	2134.54	2347.99
	0.25 B	3488.72	3732.93	4106.23
30°	0.50 B	4175.61	4467.90	4914.69
	0.75 B	5912.66	6326.55	6959.20
	1.00 B	6535.87	6993.38	7692.72
	0.00 B	4040.95	4323.81	4756.19
	0.25 B	5247.70	5615.04	6176.54
33°	0.50 B	6263.41	6701.85	7372.04
	0.75 B	7683.12	8220.94	9043.03
	1.00 B	8210.29	8785.01	9663.51
	0.00 B	5339.56	5713.33	6284.66
	0.25 B	7320.89	7833.35	8616.68

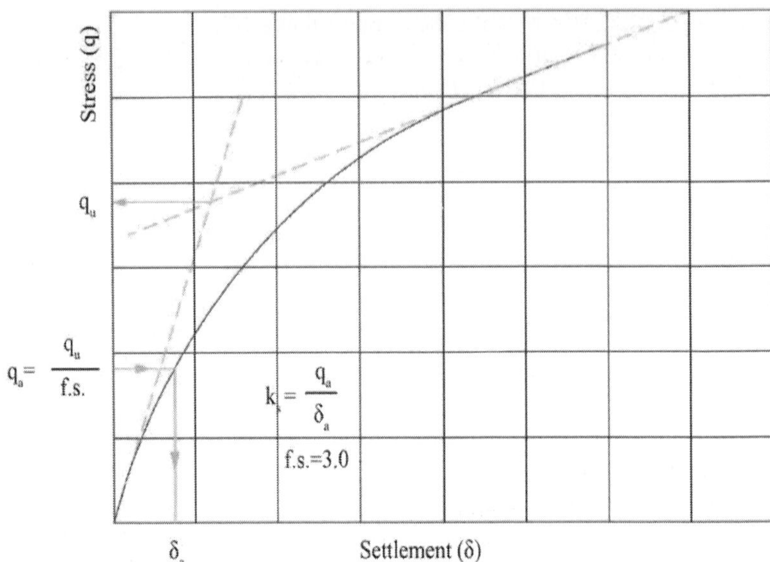

Figure 7. Determination of Subgrade Reaction "Ks".

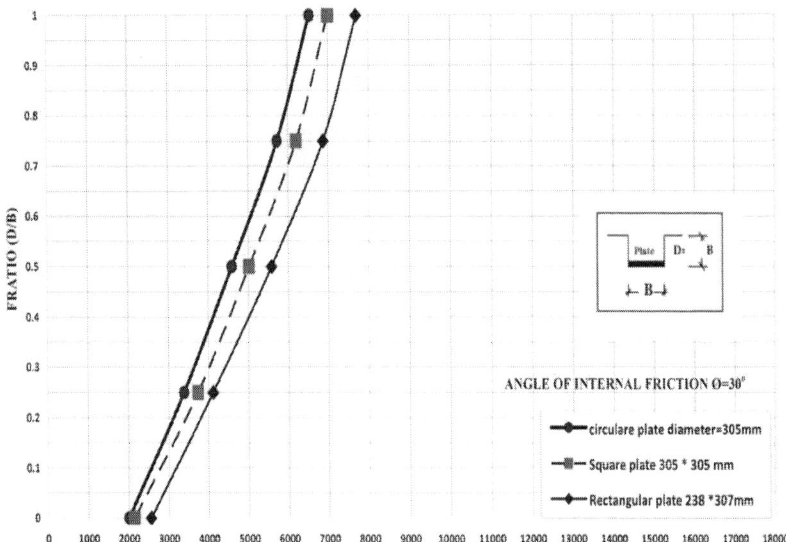

SUBGRADE REACTION K_s (kN/m^3)

Figure 8. The relationship between subgrade reaction K_s and depth of foundation for different shapes of footing for angle of internal friction Ø = 30°.

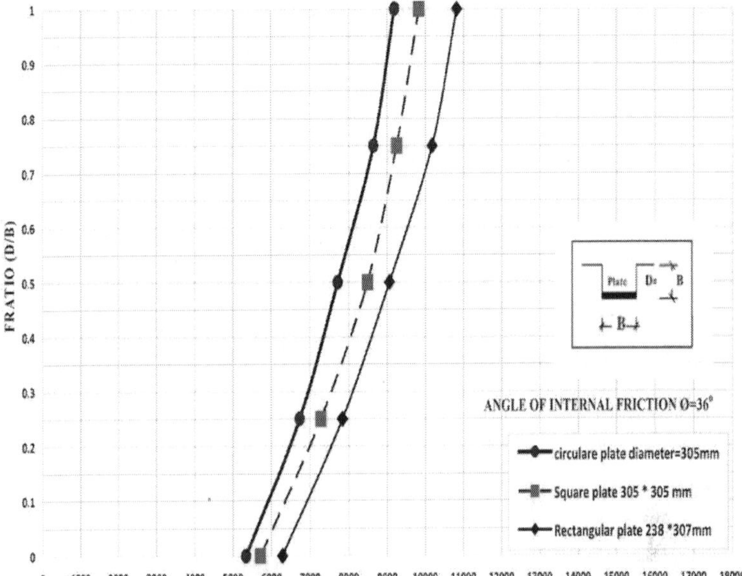

SUBGRADE REACTION K_s (kN/m^3)

Figure 9. The relationship between subgrade reaction K_s and depth of foundation for different shapes of footing for angle of internal friction $\emptyset = 36°$.

EFFECT OF FOUNDATIONS SIZE ON SUBGRADE REACTION "KS"

The relationship between subgrade reaction k_s and footing sizes for different shapes of footing has been obtained. Figures 10-12 show examples for the effect of footing sizes on k_s for different angle of internal friction. These figures show that subgrade reaction k_s of cohessionless soil increases with increasing footing size for all type of foundations. In addition, subgrade reaction k_s of cohessionless soil increases with increasing angle of internal friction.

EFFECT OF FOUNDATIONS SHAPE ON SUBGRADE REACTION "KS"

The values of subgrade reaction k_s have been obtained for different plates under different foundation shapes for different depths. Figures 13-15 give examples for the effect of footing shapes on k_s. From these fingers it can be shown that subgrade reaction k_s of cohessionless soil under rectangular plate is higher than that under square than that under circular one (at same equivalent area).

THE OBTAINED VALUES OF "K_S" FOR COHESSIONLESS SOIL

Tables 5 and 6 show the obtained value of k_s for cohessionless soil.

EMPIRICAL FORMULA

A convergence study was performed to determine the subgrade reaction (k_s) using "SPSS" statistical scientific

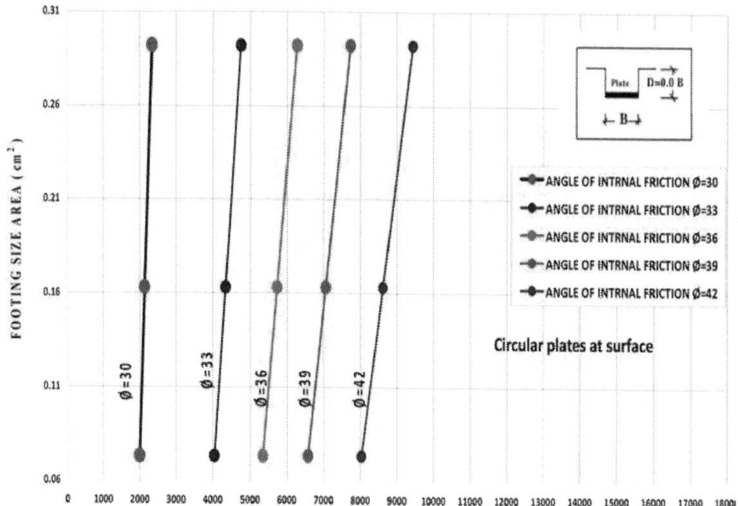

SUBGRADE REACTION K_s (kN/m³)

Figure 10. The relationship between subgrade reaction K_s and footing sizes at surface for circular plates [Size 305, 455 And 610 Mm].

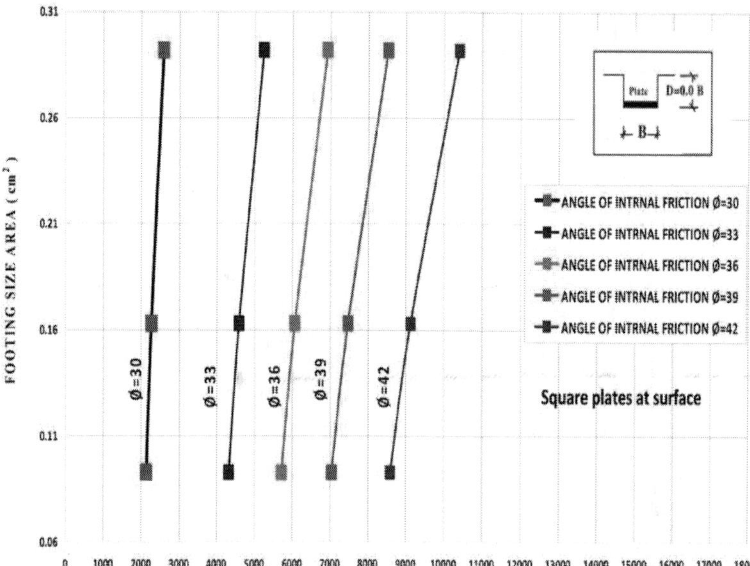

SUBGRADE REACTION K_s (kN/m³)

Figure 11. The relationship between subgrade reaction K_s and footing sizes at surface for square plates [Size Equivalent Area Of 305, 455 And 610 Mm].

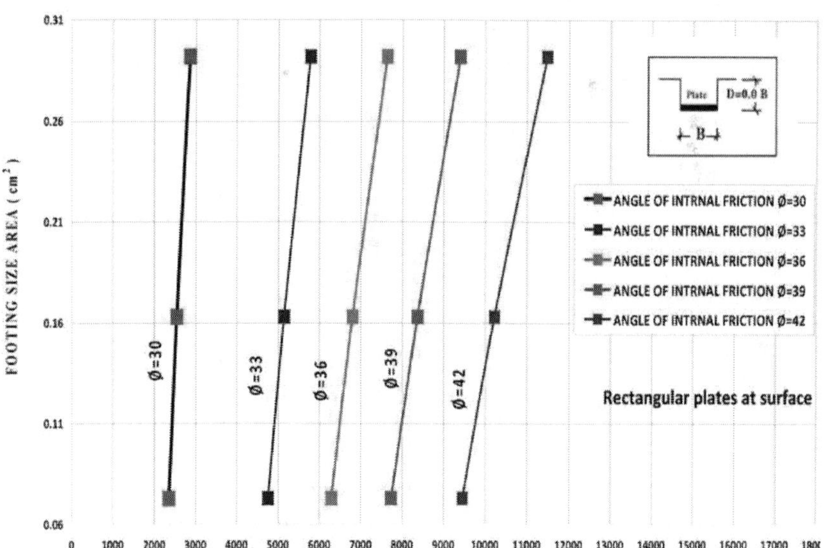

SUBGRADE REACTION K_s (kN/m³)

Figure 12. The relationship between subgrade reaction K_s and footing sizes at surface for rectangular plates [Size Equivalent Area Of 305, 455 And 610 Mm].

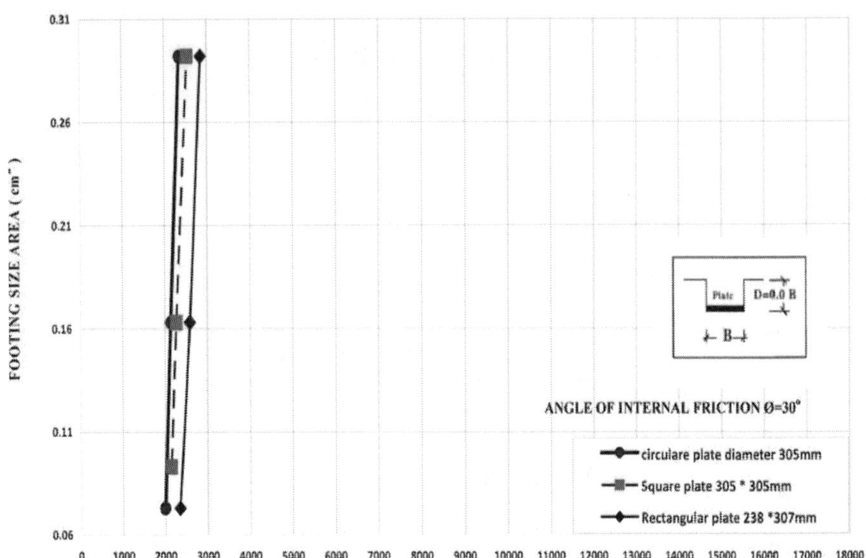

SUBGRADE REACTION K_s (kN/m³)

Figure 13. The relationship between subgrade reaction K_s and footing size at depth of footing B = 0 (At Angle Of Internal Friction Ø = 30°).

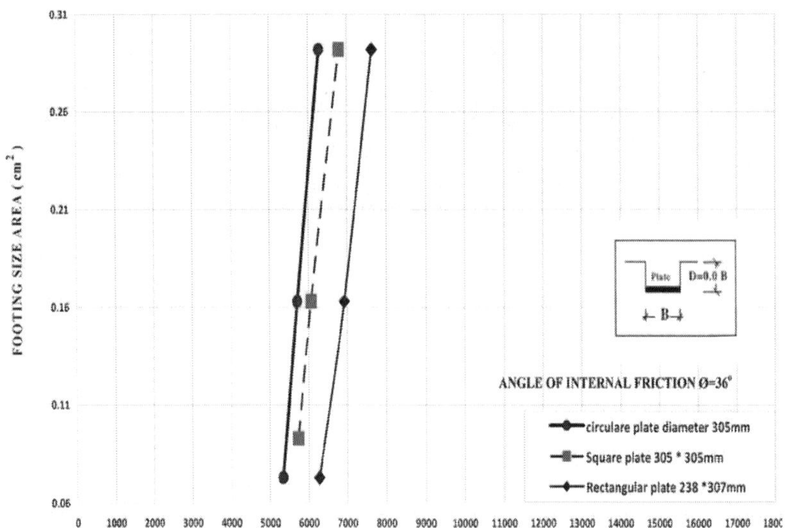

SUBGRADE REACTION K_s (kN/m³)

Figure 14. The relationship between subgrade reaction K_s and footing size at depth of footing B = 0 (At Angle Of Internal Friction Ø = 36°).

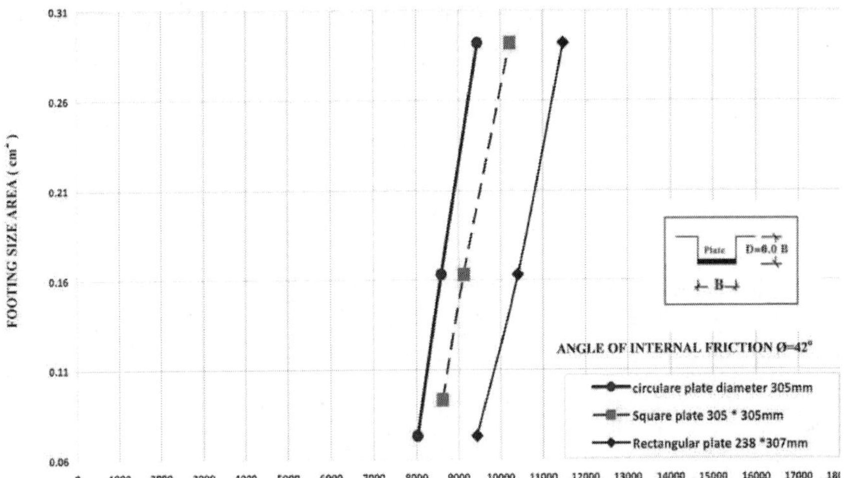

SUBGRADE REACTION K_s (kN/m³)

Figure 15. The relationship between subgrade reaction K_s and footing size at depth of footing B = 0 (At Angle Of Internal Friction Ø = 42°).

Table 5. The obtained values of "K_s" For Cohessionless Soil [MN/M³] At Depth of Foundation = 0.00 B

Relative density	"k_s" [MN/m³]		
	k_s circular plates	k_s square plates	k_s rectangular plates
Loose (Ø < 30°)	1 - 6	2 - 7	2 - 8
Medium (Ø = 30° - 36°)	5 - 9	6 - 9	6 - 10
Dense (Ø = 36° - 42°)	9 - 13	9 - 14	10 - 16

Table 6. The obtained values of "K_s" For Cohessionless Soil [MN/M³] at depth of foundation = 1.00 B

Relative density	"k_s" [MN/m³]		
	k_s circular plates	k_s square plates	k_s rectangular plates
Loose ($\varnothing < 30°$)	6 - 9	7 - 11	8 - 10
Medium ($\varnothing = 30° - 36°$)	9 - 12	9 - 13	10 - 15
Dense ($\varnothing = 36° - 42°$)	12 - 15	13 - 18	15 - 20

program. From the experimental analysis the following empirical formula is presented. The empirical formula is derived to calculate subgrade reaction (k_s) for cohessionless soil under the square foundations using regression methods:

$$k_s = \left[2529.25(D) + 290.75(\phi) + 53.68(q_a) - 170413(\delta_a) - 5881.05 \right] \tag{6}$$

where:

k_s = Modulus of subgrade reaction, (kN/m³);

$\varnothing$ = angle of internal friction;

D = depth of foundation (meter);

q_a = Allowable bearing capacity, (kN/m²);

d_a = Settlement at allowable bearing capacity (meter);

k_s is dependant and D, $\varnothing$, q_a, d_a are independent.

One-way analysis of variance (One-Way ANOVA) is used to calculate correlation coefficients of the resulting equation. The One-Way ANOVA procedure produces a one-way analysis of variance for a quantitative dependent variable by a single factor (independent) variable. Analysis of variance is used to test the hypothesis. This technique is an extension of the two-sample test. The number of cases, mean, and standard deviation, standard error of the mean, minimum, maximum, and 95%-confidence interval for the mean were calculated. It should be mentioned here that many trials were done to increase the accuracy of the derived equation (correlation coefficient = 0.95).

However, the above empirical formula gives an error % of ±10%.

COMPARISON BETWEEN VALUES OF "KS" OBTAINED FROM EMPIRICAL FORMULA AND DIFFERENT THEORETICAL METHODS

A comparison between the modulus of subgrade reaction k_s obtained by the empirical formula and the literature data mention in Table 1 has been presented in Table 7. The values of modulus of subgrade reaction "k_s" were calculated using Poisson's ratio $\upsilon = 0.3$ and the following values of Young's modulus "E_s" [kN/m^2]:

	Loose ($\emptyset < 30°$)	Medium ($\emptyset = 30° - 36°$)	Dense ($\emptyset = 36° - 42°$)
E_s [kN/m^2]	7820	14880	23080

From the above fair agreement has been obtained between values of k_s from the empirical formula at depth of footing = 0.00 B and Biot (1937) as well as Meyerhof and Baike (1965).

Table 7. The Values of "K_s" For Cohessionless Soil [MN/M^3] From The Present Work And The Literature Data.

Method	"k_s" [MN/m^3]		
	Loose ($\emptyset < 30°$)	Medium ($\emptyset = 30° - 36°$)	Dense ($\emptyset = 36° - 42°$)
Biot (1937)	2.49	5.10	8.27
Vesic (1961)	1.72	3.45	5.56
Meyerhof and Baike (1965)	2.82	5.36	8.32
Selvadurai (1984)	1.83	3.49	5.41
Author empirical formula (Depth of footing = 0.0 B)	1.56	6.17	9.00
Author empirical formula (Depth of footing = 1.0 B)	6.34	10.20	13.55

CONCLUSIONS

From the present experimental study using the in-situ plate load test (P.L.T) the followings are concluded:

1) Subgrade reaction k_s of cohessionless soil increases with increasing footing depth as well as footing size;

2) Subgrade reaction k_s of cohessionless soil under rectangular footing is higher than that under square and that under circular one (at same equivalent area);

3) Subgrade reaction k_s of cohessionless soil increases with increasing angle of internal friction;

4) The values of subgrade reaction "k_s" for cohessionless soil are presented;

5) An empirical formula is presented to calculate the subgrade reaction k_s of cohessionless soil for square foundation;

6) Fair agreement has been obtained between values of k_s obtained from the empirical formula at depth of footing = 0.00 B and Biot (1937) as well as Meyerhof and Baike (1965).

REFERENCES

1. A. B. Vesic, "Beams on Elastic Subgrade and Winkler's Hypothesis," Proceedings of the 5th International Conference on Soil Mechanics and Foundation Engineering, Paris, 1961, pp. 845-850.
2. ASTM, "Test Method for Density and Unit Weight of Soil in Place by the Rubber Balloon Method (D2167- 94)," In 1995 Annual Book of ASTM Standards, Vol. 4, No. 8, American Society for Testing and Materials, Philadelphia, 1994, pp. 167-170.
3. B. T. Iancu and O. T. Ionut, "Numerical Analyses of Plate Loading Test Numerical Analyses of Plate Loading Test," Buletinul Institutului Politehnic Din IASI Publicat de Universitatea Tehnică, Gheorghe Asachi," Tomul LV (LIX), Fasc. 1, Sectia Construct II. Arhitectură (IancuBogdan Teodoru and Ionut-Ovidiu Toma), 2009, pp. 57-65.
4. E. Winkler, "Die Lehre von Elastizitat and Festigkeit (on Elasticity and Fixity)," Praguc, 1987, p. 182.
5. Egyptian Code, "Soil Mechanics and Foundation," Organization, Cairo, 2001.
6. J. E. Bowles, "Foundation Analysis and Design," 6th Edition, McGrow-Hill International Press, 1998.
7. K. Terzaghi, "Evaluation of Coeffcients of Subgrade Reaction," Géotechnique, Vol. 5, No. 4, 1955, pp. 297-326. http://dx.doi.org/10.1680/geot.1955.5.4.297
8. L. S. Ping and L. W. Yang, "Subgrade Reaction and Load-Settlement Characteristics of Gravelly Cobble Deposits by Plate-Load Tests," Canadian geotechnical Journal, Vol. 35, No. 5, 1998, pp. 801-810.
9. M. A. Biot, "Bending of Infinite Beams on an Elastic Foundation," Journal of Applied Mechanics Trans. Am. Soc. Mech. Eng., Vol. 59, 1937, pp. A1-A7.
10. M. Aminaton, L. Nima, J. Masoud, Kh. Mehrdad, Kh. Mahdy, A. Payman and D. B. Ali, "Foundation Size Effect on Modulus of Subgrade Reaction on Sandy Soils," The Electronic Journal of Geotechnical Engineering, Vol. 17, 2012, pp. 2523-2530.
11. M. K. Elsamny, M. B. Elsedeek and W. N. Abd Elsamee, "Effect of Depth of Foundation on Modulus of Elasticity 'Es' for Cohessionless Soil," Civil

Engineering Researches Magazine of Al-Azhar University, Vol. 32, No. 3, 2010, p. 938.

12. S. K. Dae and Y. P. Seong, "Relationship between the Subgrade Reaction Modulus and the Strain Modulus Obtained Using a Plate Loading Test," 9th WCRR Lille World Congress, 2011.

13. Z. M. Reza and J. Masoud, "Foundation Size Effect on Modulus of Subgrade Reaction in Clayey Soil," The Electronic Journal of Geotechnical Engineering, Vol. 13, 2008, pp. 1-8.

CITATION

Elsamee, W. (2013) An Experimental Study on the Effect of Foundation Depth, Size and Shape on Subgrade Reaction of Cohessionless Soil. *Engineering*, **5**, 785-795. doi: 10.4236/eng.2013.510095.

CHAPTER 7

Geochemical Modeling of Trivalent Chromium Migration in Saline-Sodic Soil During Lasagna Process: Impact on Soil Physicochemical Properties

Salihu Lukman[1], Alaadin Bukhari[2], Muhammad H. Al-Malack[2], Nuhu D. Mu'azu[3], and Mohammed H. Essa[2]

[1]Department of Civil Engineering, ACHB, King Fahd University of Petroleum and Minerals, Hafar Al-Batin 31991, Saudi Arabia
[2]Department of Environmental Engineering, University of Dammam, Dammam, Saudi Arabia
[3]Environmental Engineering Department, University of Dammam, Dammam 31451, Saudi Arabia

ABSTRACT

Trivalent Cr is one of the heavy metals that are difficult to be removed from soil using electrokinetic study because of its geochemical properties. High buffering capacity soil is expected to reduce the mobility of the trivalent Cr and subsequently reduce the remedial efficiency thereby complicating the remediation process. In this study, geochemical modeling and migration of trivalent Cr in saline-sodic soil (high buffering capacity and alkaline) during integrated electrokinetics-adsorption remediation, called the Lasagna process, were investigated. The remedial efficiency of trivalent Cr in addition to the impacts of the Lasagna process on the physicochemical properties of the soil was studied. Box-Behnken design was used to study the interaction effects of voltage gradient, initial contaminant concentration, and polarity reversal rate on the soil pH, electroosmotic volume, soil electrical conductivity, current, and remedial efficiency of trivalent Cr in saline-sodic soil that was artificially spiked with Cr, Cu, Cd, Pb, Hg, phenol, and kerosene. Overall desirability of 0.715 was attained at the following optimal conditions: voltage gradient 0.36 V/cm; polarity reversal rate 17.63 hr; soil pH 10.0. Under these conditions, the expected trivalent Cr remedial efficiency is 64.75 %.

INTRODUCTION

In early 1992, a discussion took place between the then Monsanto Chief Executive Officer (CEO) and Administrator of the United States Environmental Protection Agency (USEPA) which ultimately led to the invention of the Lasagna process [1]. In the late 1993, Brodsky and Ho of Monsanto filed the first Lasagna U.S. patent followed by a second one, all published in 1995 [2, 3]. In the Lasagna process, contaminated soil is remediated by creating at least one liquid permeable zone within a contaminated soil region and turning it into treatment zone. Appropriate materials (sorbents, catalytic agents, microbes, oxidants, and buffers) are then introduced into the treatment zone. An electrode is placed at the first end of the contaminated soil region and another of opposite charge is placed at the opposite end of the contaminated soil region. A direct electric current is then transmitted through the contaminated soil region between the two electrodes. This causes movement of water and dissolved organic and inorganic materials in subsurface soils from one electrode (anode) to the other (cathode) under electroosmosis as a result of current movement from anode to cathode. In 1802 electroosmosis was first observed; detailed study of the mechanism was done by Reuss [4] in his classic experiment reported by Abramson [5]. In 1909, Freundlich and Neumann [6] provided the general name "electrokinetic phenomena" to refer to the electrically driven mass flow of dissolved contaminants and pore fluid transport in soils induced by an applied DC voltage. It is made up of transport of pore fluid via electro osmosis (EO) and transport of ions or charged species via electromigration [7]. The direction and quantity of contaminant movement are influenced by the contaminant concentration, solubility, speciation, degree of hydrophobicity, soil type and structure, and the mobility of contaminant ions, as well as the interfacial chemistry and the conductivity of the soil pore fluid [8]. The remedial efficiency generally depends on the nature of the contaminants and soil properties, such as pH, permeability, adsorption capacity, buffering capacity, and geochemical processes (such as acid/base reactions and migration, dissolution/precipitation, redox reactions, complexation, and speciation) [7, 9]. Saline-sodic soils possess electrical conductivity above 4 dS/m, soil paste pH greater than 8.2, and exchangeable sodium percentage greater than 15 [10, 11].

First application of electrokinetics took place in India in the 1930s. It was used to remove excess salts from alkali soils in order to restore it to arable condition [13].

Following its invention in 1993, extensive studies started in 1994 in bench-scale [14] then scaled up in a pilot-scale under laboratory conditions. The first field test called "Phase I: Small Field Test" was conducted in 1995 at the Paducah gaseous diffusion plant (PGDP) site whose soil was contaminated with trichloroethylene (TCE). Full-scale remediation using the Lasagna technology was undertaken at two other contaminated sites in the United States [15]. The specific details of all the Lasagna process implementations are presented in Tables 1 and 2. It is noteworthy that only two of the studies have considered simultaneous removal of contaminant mixture. All others have dealt with only a single organic compound or heavy metal. It has already been observed that contaminated soils do not contain single contaminants only but usually several pollutants appear in the soil in mixed components [16–19].

Table 1. Applications of Lasagna process at bench-scales from inception to date.

Treatment zone material	Contaminant	Soil type	Cell dimensions (length × width × depth)	Polarity reversal/ downtime	Removal efficiency, %	Voltage gradient, (V/cm)/ current (mA)	Power consumption, kWhr/m^3	Run time, days	Electroosmotic conductivity, cm^2 V^{-1} s^{-1} (×10^{-5})	Treatment zone spacing, cm	Reference
AC* + sand, bacteria + AC + sawdust	p-nitrophenol	Kaolinite	10 cm ID, 21.6 cm long	Yes/continuous	90–99	1–7/3 (constant)	10	20	2.5	6	[14, 20]
AC (Bamboo	Cd	Sandy loam	24 cm × 10 cm ×	Yes/continuous	79.6	1/7–27	—	12	—	10	[21]
AC (Bamboo	Cd	Kaolin	24 cm × 10 cm ×	No/continuous	93	1/3–23	—	8	—	10	[22]
AC (Bamboo charcoal)	2,4-dichlorophenol and	Sandy loam	24 cm × 10 cm × 10 cm	Yes/continuous	75.97 (Cd);54.92 (2,4-	1/Variable	121.91–128.48	10.5	—	16	[23]
GAC**	Cr, Cd, Cu, Pb, Hg, Zn, phenol, kerosene	Saline-sodic clay	24 cm × 10 cm × 12 cm	No/continuous	75.9 (Cr); 34.4 (Cd); 41 (Cu); 55.8 (Pb); 92.49 (Hg); 26.8 (Zn); 100 (phenol); 49.8 (kerosene)	0.6–1/880	1777–4273	21	425	6	[24]

*Activated carbon; **Granular activated carbon.

Table 2. Applications of Lasagna process at pilot- and field-scales from inception to date.

Treatment zone material	Contaminant	Soil type	Site dimensions (length × width × depth)	Polarity reversal /downtime	Removal efficiency, %	Voltage gradient, (V/cm)/ Current (A)	Power consumption, kwh/m^3	Run time, month	Electroosmotic conductivity, $cm^2\,V^{-1}\,s^{-1}$ (×10^{-5})	Treatment zone spacing, cm	Reference
AC + sand	p-nitrophenol	Kaolin/ Kaolinite/clay loam	1.22 m × 0.61 m × 0.61 m	Yes/ continuous	98	1 (constant)/96.2 (based on current density)	51	3	0.56–1.7	35.6	[20]
GAC[1]	TCE[2]	Clay loam	4.6 m × 3 m × 4.6 m	Yes/continuous	99	0.35–0.45/40 (constant)	—	4	1.2	60	[25]
Iron filings + kaolin	TCE	Clay loam	6.4 m × 9.2 m × 13.7 m	Yes/3-week	95–99	0.23–0.31 (constant)/110–200	—	12	1.2	60 & 150	[26]
Iron filings + kaolin	TCE	Clay loam	27.4 m × 22 m × 13.5 m	No/pulse mode	99	0.15–0.26 (constant)/500–700	—	24	—	150	[15, 27]
Iron filings + kaolin	TCE	Clay loam	33 m × 24 m × 7.5 m	—	60 (after 1 year)	0.16 (constant)/250–400	—	24	—	150	[15]

[1]Granular activated carbon; [2]Trichloroethylene

The geochemical properties of the most stable forms of Cr, that is, trivalent and hexavalent Cr under electrokinetic remediation, have been extensively studied in different types of soils (kaolin, glacial till, etc.) by Reddy and his coworkers and other investigators [28–36]. The trivalent Cr, though considered relatively nontoxic compared to the hexavalent Cr, exists in the subsurface environments as cation, Cr^{3+}, and in the following hydroxocomplex forms: $Cr(OH)_4^-$, $CrOH^{2+}$, and.$Cr(OH)_3^0$.Cr^{3+} and $CrOH^{2+}$ ions are mostly prevalent at soil pH values less than 6, while $Cr(OH)_4^-$ and $Cr(OH)_3^0$ ions prevail when pH is greater than 11.8. The redox state also affects the Cr state with reduced state favoring the presence of the trivalent Cr while the oxidized state favors the existence of the hexavalent Cr. Most of the trivalent Cr species are less mobile because of their low solubility over wide pH range (<12) and may be readily adsorbed by the negatively charged clay surfaces. There exists redox state in the subsurface environment because of the generation of oxygen and hydrogen gases at the electrodes in addition to the possible presence of iron (reducing agent), manganese (oxidizing agent), or microorganisms. The redox potential (Eh) and soil pH determine the possible oxidation of Cr from the trivalent to hexavalent form as shown in Figure 1. Chinthamreddy and Reddy [28] have found no significant oxidation of trivalent Cr in high buffering capacity soil such as glacial till.

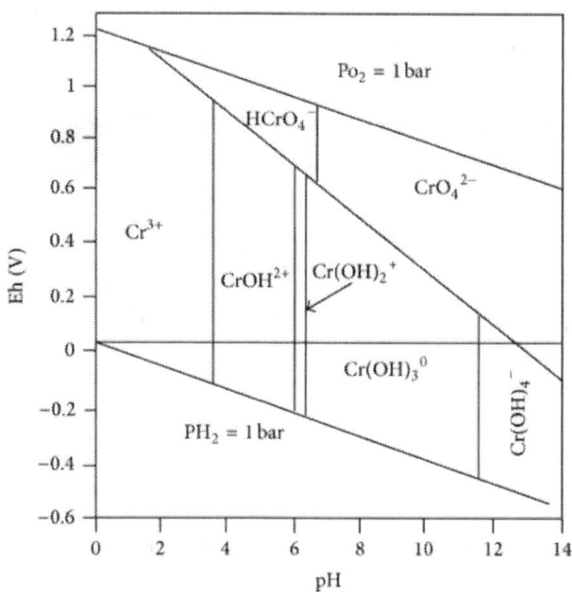

Figure 1. Redox potential (Eh)-pH diagram for Cr–O–H system [12].

Empirical modeling using response surface methodology (RSM) offers great and numerous advantages which include large amount of information from a small number of experiments, evaluation of simultaneous interaction effects of the independent parameters on the responses, and simultaneous optimization of multiple factors and responses for obtaining optimal conditions [37,38]. The key success of RSM is uncovering interactions of factors which cannot be achieved using the traditional one-factor-at-a-time (OFAT) optimization approach [39]. Fundamental understanding of the physics and chemistry which governs the process is essential in determining the influential factors to be investigated and their levels or ranges are necessary for successful implementation of RSM for any process modeling and optimization. Basically, there exist four different experimental designs for RSM implementation: 3-level factorial design (3FD), Box-Behnken design (BBD), central composite design (CCD), and Doehlert design (DD). Bezerra et al. [38] have reviewed each of these design methods. The Box-Behnken design is obtained by combining two-level factorial designs with incomplete block designs followed by adding a specified number of replicated center points. BBD is preferred when investigating three factors (3) using RSM, because it will give enough information for analyzing factor-response interactions from the least experimental runs when compared to 3FD and CCD.

Some microbially-driven biotransformation processes may affect the soil physicochemical properties after electrokinetic remediation because of the passage of electric current and development of pH gradients [40]. These lead to original soil mineral degradation and alteration via biotransformation. While biotransformation deals with the bioweathering and alteration or degradation of clay minerals, biomineralization refers to the formation of amorphous and crystalline materials from aqueous ions by biologically mediated processes. In addition to current and pH gradients, heavy metals also cause to affect the following biological assays: soil microbial biomass carbon, enzyme activity, basal soil respiration, and earthworm assays and seed assays [41–44]. Given the aforementioned intricacies of the geochemical behavior and migration of trivalent Cr in soil during electrokinetic remediation, this study was aimed at investigating trivalent Cr migration and remedial efficiency in high buffering capacity and alkaline soil during electrokinetic study in addition to the impacts of the soil remediation on the physicochemical properties of the soil. A carefully designed experiment using BBD was used to study the interaction effects of voltage gradient, initial contaminant concentration, and polarity reversal rate on the trivalent Cr remedial efficiency in saline-sodic soil that was artificially spiked with Cr, Cu, Cd, Pb, Hg, phenol, and kerosene using RSM modeling and optimization tools.

MATERIALS AND METHODS

Characterization

Natural saline-sodic clay, obtained from Al-Hassa Oasis, Saudi Arabia, was used in this study. The soil has the following characteristics: pH (8.3), moisture content (3.91%), soil organic matter (2.59%), electrical conductivity (15.24 dS/m), specific surface area (9.07 m^2/g), pore volume (0.014 cm^3/g), pore size (62.55 Å)—mineralogy from X-ray diffraction (XRD), quartz (SiO_2) (87.4%), calcite ($CaCO_3$) (5.2%), and dolomite ($CaMg(CO_3)_2$) (7.4%). X-ray fluorescence spectroscopy (XRF) revealed that the soil consists of the following elements: Ca (37.64%), Si (34.73%), Fe (10.41%), Al (7.6%), K (3.42%), Mg (2.48%), Pd (2.85%), and Ti (0.86%). These properties were determined using methods of the American Society of Testing and Materials (ASTM) standards and were reported elsewhere [45]. The granular activated carbon (GAC) used in the present study whose surface area is 952 m^2/g was produced locally from date palm pits using phosphoric acid impregnation method. Its characterization and properties have been reported elsewhere [46, 47].

Adsorption Testing

Single and competitive adsorption of five heavy metals (Cr, Cd, Cu, Zn, and Pb) were performed to determine the selectivity sequence and to understand the adsorption behavior of these metals under different pH conditions. This is particularly important to this study, because soil mineralogy affects heavy metal adsorption behavior and selectivity sequence under different pH conditions. Lukman et al. [45] reported the detailed procedures carried out for the competitive adsorption testing.

Coupled Electrokinetics-adsorption Study

Fifteen (15) bench-scale experiments, each having a 21-day run time, were designed and performed to investigate the migration and distribution of trivalent Cr in a contaminant mixture using the coupled electro kinetics-adsorption technique and to understand the operating variables' effects on saline-sodic soil.

Reactor Design and Experimental Procedures

The Plexiglas reactor total volume was about $2268\,cm^3$, made of seven chambers. The overall reactor dimensions are 24 cm (long) $\times$ 10 cm (width) $\times$ 12 cm (depth). Approximately 1 kg of local KSA soil was artificially spiked with kerosene, heavy metals (Cu, Cr, Cd, Pb, Zn, and Hg), and phenol at predetermined concentrations. Thorough mixing was done using mechanical mixer (Gilson Company Inc.) so as to achieve a homogeneous distribution of the contaminants around the soil matrix. The mixed spiked soil was placed in a fume-hood for drying over a period of time necessary to evaporate the solvents (hexane and distilled water). Distilled water was added to adjust the final moisture content of the soil to about 33–70%. The initial conditions of the soil pH, moisture content, organic matter, and electrical conductivity were measured as well as the actual initial concentrations of the contaminants. Then, the uniformly mixed contaminated soil was placed into the cell layer by layer. Each layer was compacted with stainless steel spatula so that the amount of void space was minimized. The reactor used for the experiments consists of the cell, two graphite electrodes serving as anode and cathode, DC power supply (LG, GP-505), processing fluid reservoirs, heavy duty recirculation pump (BVP Instratec), portable data logger (TDS-303, Tokyo Sokki Kenkyujo Co., Ltd) for real-time monitoring of temperature, electric current, and voltage across the system (Figure 2). The two electrode compartments with 240 mL working volume, placed at each end of the cell, were isolated from the soil zone by a porous Perspex plate and filter paper. The conditioning of the electrolyte was controlled using anolyte (2N NaOH) and catholyte (1N HNO_3). The pH of the processing fluids was monitored every 8 hr for the 21-day duration of each test. Based on the pH and volume the processing fluids remaining in the electrode chambers, complete replacement, or refill were carried out accordingly. Two planar-shaped electrodes, 10 cm $\times$ 10 cm $\times$ 0.5 cm, were used to generate a uniform electric field. Within the described cell, two treatment zones that cut across the cell vertically bracketing the spiked soil compartment were filled with the GAC. The data monitoring system was recording electric current variation, applied voltage, and temperature of the soil compartments online following a 30 min preset time step and automatically stores them for subsequent retrieval using floppy disc which can be read using personal computer for easy data and energy consumption analysis. The power supply provides a constant DC electric voltage for the electrokinetic tests. Every week, fractions of the soil specimens were taken at the center of each chamber to determine the residual concentrations of the contaminants, soil pH, water content, organic matter, and electrical conductivity. Upon the completion of each test, the electrode assemblies were disconnected and the soil specimen was extruded from the cell, sectioned into parts, weighed, and preserved in glass vials for organic extraction, heavy metal digestion, and subsequent analyses using the analytical procedures outlined below.

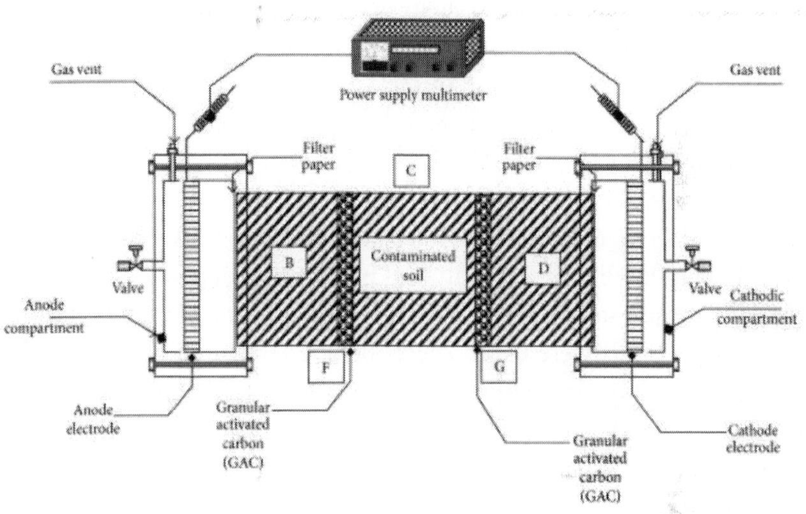

Figure 2. Coupled electrokinetics-adsorption experimental setup.

Analytical Procedures for Contaminant Extraction and Analysis
Heavy Metals

Extraction of heavy metals from soil samples was performed according to guidelines spelt out in EPA method 3050B for acid digestion of soils, sediments, and sludges [48] and analyzed using flame atomic absorption spectrometry (AAnalyst 700, Perkin Elmer). All soil samples were extracted in duplicate. EPA method 7000B [49] was employed for heavy metal analysis using flame atomic absorption spectrometry except for mercury which was analyzed using mercury analyzer (Solid Mercury Analyzer SMS 100, Perkin Elmer) according to EPA method 7473 [50].

Visual MINTEQ 3.0 [51] was employed to model the metal ion speciation using its dissolved concentration, pH, temperature, and ionic strength.

Kerosene and Phenol

A mixture of methylene chloride and hexane (1 : 1) (v/v) was used as the extraction solvent. Soil samples were extracted using pressurized fluid extraction according to EPA method 3545 procedures [52] using accelerated solvent extractor (ASE 200, Dionex). Volume of extract generated was then injected into the GC-MS (Clarus 580, Perkin Elmer) equipped with autosampler for analysis. TPH quantification was done by using the total chromatographic area counts using retention time range for the elution of hydrocarbon within the kerosene range C_8–C_{16}. Guidelines spelt out in EPA mMethod 8270D [53] for the quantification of semivolatile organics by GC-MS were adhered to.

Data Reliability: QC Protocols, Accuracy, and Precision

To evaluate reliability of the analytical procedures, duplicate samples were analyzed for each sample. Quality control (QC) protocols spelt out in EPA method 7000B [49] were used. These include the use of initial calibration blank (ICB), initial calibration verification (ICV), continuous calibration verification (CCV), and continuous calibration blank (CCB). The accuracy of the spiked soil samples was evaluated using percent recovery set at about ±30% of the spiked value [49]. Repeatability of the experimental results was assessed by ensuring that the precision obtained using the relative percent difference (RPD) was not above 30%.

Testing Program and Mathematical Model Development

Box-Behnken design (BBD) was chosen for the experimental design because of its advantages over central composite design (CCD) and 3-level factorial design when dealing with only three factors. In BBD, the experimental points are hyperspherically arranged, equidistant from the central point [38]. Response surface methodology (RSM) was used in modeling, optimization, and interpretation of the results with the help of Design-Expert version 8 (Stat-Ease, Inc.) platform [39, 54]. The investigated variables (called factors in RSM) are the polarity reversal rate, voltage gradient, and initial contaminant concentration designated as A, B, and C, respectively. These variables were selected based on their known influence on contaminant remedial efficiency and were coded and varied according to Table 3. Based on the factor levels and the chosen number of central points (3), a total of fifteen (15) experiments were randomly designed, using the BBD (Table 4), and subsequently conducted. Only one central point is shown in Table 4.

Table 3. Codification and ranges of factors

Variable	Designation	Units	Coded variable levels		
			−1	0	1
Polarity reversal	A	hr	0	24	48
Voltage gradient	B	V/cm	0.2	0.6	1
Concentration	C	mg/kg	20	60	100

Table 4. Design of experimental runs using the Box-Behnken design

Run order	Polarity reversal, (hr)	Voltage gradient, B(V/cm)	Concentration, C(mg/kg)	Remedial efficiency, %
1	0	0.6	20	0
2	48	0.6	20	0
3	24	1	20	0
4	24	1	100	0
5	24	0.6	60	79.97
6	0	1	60	72.73
7	24	0.2	20	0
8	0	0.2	60	36.93
9	48	1	60	65.66
10	48	0.6	100	0
11	0	0.6	100	25.5
12	24	0.2	100	0
13	48	0.2	60	34.88

This experimental design was preliminarily evaluated using variance inflation factor (VIF) to check for orthogonality (independence of factors) and leverage which quantitatively measures the potential of a design point to have significant influence on model fit [54]. These were determined using (1) and (2), respectively. VIF value of 1 indicates that the factor is orthogonal to all other factors in the design. In a case whereby factors are highly correlated, then, R^2 value becomes a poor indicator of model's predictive ability and it becomes more and more difficult to unravel how each of the investigated factors affect the response. Experimental points having high leverage values close to 1 should influence the model fit by carrying any error (experimental or measurement) into the model; as such, they should be conducted more carefully [39]:

$$\text{VIF} = \frac{1}{\left(1 - R_i^2\right)}$$

(1)

$$\text{Leverage} = \frac{p}{n},$$

(2)

Where R_i^2 is the coefficient of determination; p is the number of model terms; and n is the number of experiments.

Following design evaluation, the responses were fitted to a quadratic model which was fine-tuned by removing any insignificant term. This will maximize R^2 and minimize lack of fit. The general quadratic equation for fitting models in RSM is

$$y = \beta_o + \sum_{i=1}^{k} \beta_i x_i + \sum_{i=1}^{k} \beta_{ii} x_i^2 + \sum_{1 \leq i \leq j}^{k} \beta_{ij} x_i x_j + \varepsilon,$$

(3)

where y is the response or dependent variable; k is the number of factors; β_o, β_i, β_{ii}, and β_{ij} are the coefficients to be fitted using regression for constant term, linear, quadratic, and interaction parameters, respectively; and x is the variables.

The developed models were evaluated using the rich diagnostic tools provided in Design-Expert which include normal plot of residuals (to test the assumption of normality of residuals), predicted versus actual plot (to test the assumption of constant variance), Box-Cox plot (to check the need for data transformation), and externally studentized residuals (to check the presence of any outlier in the data). The effects of factors were compared at a particular point in the design space using the perturbation plot. Response surface and contour plots were then generated.

Use of Desirability Function in Numerical Optimization
Desirability function, being one of the mathematical methods for computation of critical values (maximum or minimum) and measuring overall success of optimizing multiple responses using geometric mean, was employed for the optimization of trivalent Cr remedial efficiency. A search for a combination of factor levels which simultaneously satisfies the goals imposed on factors and responses is first carried out, followed by combining these goals into an overall desirability function that ranges from 0 (outside of the optimization limits) to 1 (at the goal). Combining all responses into overall desirability eliminates favoring one response over another. The aim is not to clinch a desirability value of 1 but to find a good set of conditions that will meet all the set goals for each factor and response [55]. This is achieved using numerical optimization algorithms [52].

RESULTS AND DISCUSSION

Discussion of the monitored results obtained after performing thirteen (13) tests with 3 centre points will be focused on geochemical processes affecting sorption/desorption and migration/removal mechanisms such as the development of acid/base fronts, migration and reactions, dissolution/precipitation, oxidation/reduction reactions, complexation, and metallic ion speciation. In addition, presentation of the developed mathematical models and discussion on how the factors affect the respective responses will follow.

Single and Competitive Adsorption of Heavy Metals on Clay

Lukman et al. [24, 45] have discussed the physicochemical characteristics of the saline-sodic soil. Additional discussion will be provided in the subsequent sections. Lukman et al. [45] have found out that the adsorptive capacities of Cu and Zn ions are higher in the multicomponent adsorption scenario than in the single component scenario. The adsorption selectivity sequences obtained using the coefficient of distribution for the single and multicomponent scenarios are Cr > Pb > Cu > Cd > Zn and Cr > Cu > Pb > Cd > Zn, respectively [45]. Yong et al. [40] have identified the general factors that influence selectivity sequence to be ionic size or activity, first hydrolysis constant, soil type, and pH of the system. From the multicomponent desorption study, trivalent Cr ions were tightly held by the soil surface, thus having the least percentage desorption, followed by Cd and Cu ions. Reddy and his coworkers [28, 30, 32] have reported that trivalent Cr ions adsorb highly to soil solids and form cationic species that are insoluble over a wide range of pH. This is in line with the present findings by Lukman et al. [45] which revealed high selectivity for the trivalent Cr during multicomponent adsorption and desorption tests.

Soil pH Distribution, Electrical Conductivity, Bipolar Effects, Electroosmotic Flow, and Current

Soil pH and Electrical Conductivity

The soil pH (8.3) indicates that it contains appreciable soluble salts capable of undergoing alkaline hydrolysis such as sodium carbonate [11]. The hydrolysis of calcite and dolomite may be limited by their low solubility, thus producing a pH of about 8–8.2 in soils. In addition, Na^+ ions do not strongly compete with H^+ ions for exchange sites as do Ca^{2+} ions that are strongly and more tightly held on the soil surface. The inability of the displaced Na^+ ions to inactivate OH^- ions results in increased soil pH, which is usually greater than 8.2. Moreover, for a soil whose pH is greater than 8.2, its exchangeable sodium percentage has to be greater than 15 [11]. Presence of calcite and dolomite coupled with alkaline hydrolysis of sodium carbonate gives high electrical conductivity to the soil (15.24 dS/m).

The saline-sodic nature of the soil necessitates the use of processing fluids (2N NaOH and 1N HNO_3) to continuously neutralize the rapidly generated

H^+ and OH^- ions at the anode and cathode, respectively. These fluids were monitored every 8 hours and replaced as they degraded. HNO_3 and $NaOH$ are strong acid and base, respectively, and dissociate completely according to the following reactions:

$$HNO_3\,(l) \longrightarrow H^+\,(aq) + NO_3^-\,(aq)$$

$$(4)$$

$$NaOH\,(aq) \longrightarrow Na^+\,(aq) + OH^-\,(aq).$$

$$(5)$$

Because of the electrochemical decomposition of water, OH^- and H^+ ions are produced at the cathode and anode, respectively, as shown in (6) and (7):

$$4H_2O\,(l) + 4e^- \longrightarrow H_2\,(g) + 4OH^-\,(aq)$$

$$(6)$$

$$2H_2O\,(l) \longrightarrow O_2\,(g) + 4H^+\,(aq) + 4e^-.$$

$$(7)$$

The electrochemically generated H^+ and OH^- ions due to water electrolysis at the anode and cathode, respectively, are neutralized to form water molecules (8) because of the OH^- and H^+ ions produced from the dissociation of the catholyte and anolyte, respectively, as shown in (5) and (4):

$$H^+\,(aq) + OH^-\,(aq) \longrightarrow H_2O\,(l).$$

$$(8)$$

The oxygen and hydrogen gases generated may be vented out, while some amount may go into the soil and alter the redox chemistry [56]. Na^+ and NO_3^- ions migrate into the soil to the opposite electrodes thereby increasing the electrical conductivity as the treatment process progresses. A sustained and variable electroosmotic flow was observed due to the migration of the Na^+ ions, which could enhance the migration of the double layer complexes toward the cathode, while nitrate ions could be involved in complex formation with the cations [28]. This electroosmotic flow will lead to decreasing volume of the anolyte and increasing volume of the catholyte over time. Hence, refilling the anolyte is necessary if it has not degraded completely. In addition, since the processing fluids are finite in volume and the electrochemical decomposition of water at the electrodes is continuous for the test duration, then, a time will be reached when all the ions in the processing fluids have been exhausted. Consequently, rise and fall in catholyte pH and anolyte pH, respectively, are expected before the complete replacement of the processing fluids. Now, ions generated at the cathode according to (6) migrate into soil toward the anode. In this migration process, soil pH rises (Figure 3) and metal hydroxides are formed which could precipitate and reduce the electrical conductivity (Figure 4) and increase current consumption near the cathode [41]. At the same time, soluble hydroxocomplexes are formed with the cations due to complexing property of the hydroxyl ions [24, 57]. On the other hand, hydrogen ions generated at the anode (7) migrate toward the cathode. This process may lead to soil protonation or desorption of indigenous and spiked heavy metals and hence increase the electrical conductivity (Figure 4) [32]. Given the presence of calcite and dolomite in the soil minerals, the developing acid front may be buffered by the carbonate mineral, thereby hindering any fall in the soil pH (Figure 3). From the forgoing discussion, it is clear that there will be an overall increase in the soil pH and electrical conductivity (Figure 4) as the integrated electrokinetics-adsorption remediation progresses. Results obtained for electrokinetic remediation of high buffering capacity glacial till by Reddy and hiscoworkers [30–32, 58] have corroborated these findings. The transient nature of the acid/base front migration and reactions may be responsible for the lower final values of some pH and EC than the preceding 1st or 2nd week values. In addition, the electroosmotic flow (Figure 5) will undoubtedly vary spatially and temporally as it also depends on the soil zeta potential, processing fluids pH, pore fluid viscosity, and permittivity [7, 59–61].

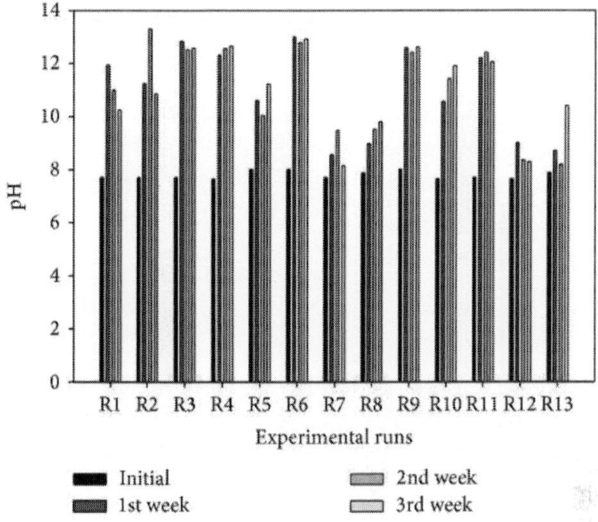

Figure 3. Weekly pH variation.

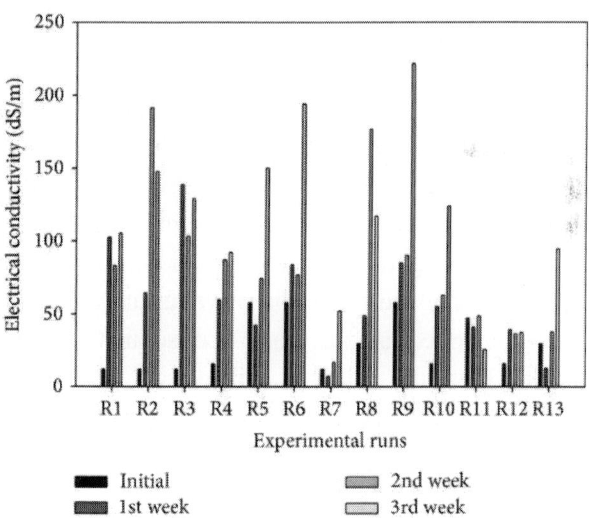

Figure 4. Weekly soil electrical conductivity variation.

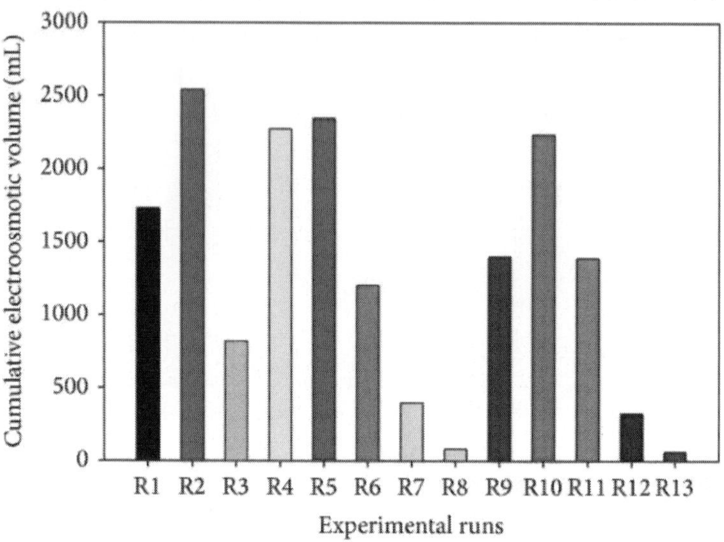

Figure 5. Cumulative electroosmotic volume for each test.

It was observed from Figure 3 that the average initial soil pH after spiking is within the range 7.7–8, lower than the original soil pH (8.3), while the final pH ranges from 8 to 12.9. The lower initial pH was due to the acidity of the contaminant solutions, while the higher final pH values resulted from the high buffering capacity of the soil which neutralized the generated acidic front from (7) but allowed the migration of the basic front generated from (6). In addition, the hydroxyl ions generated from the dissociation of NaOH (5) and reduction of water at the cathode (6) aid in neutralizing the generated acidic front. Consequently, all weekly pH values are higher than the initially spiked soil pH for all the tests. Additionally, low pH rise (8–10.4) was observed for all the tests conducted using 0.2 V/cm (R7-8, R12-13) whereas highest pH (12.6–12.9) was recorded for all tests conducted using 1 V/cm (R3-4, R6, and R9) consistently. High voltage gradient leads to the passage of high amount of current which increases the rate of the electrochemical decomposition of the electrolyte and enhances subsequent migration of the basic front into the soil. This basic front migration is responsible for raising the soil pH. This observed effect of the voltage gradient on the soil pH has been successfully modeled mathematically and the coded linear model equation at 5% significant level (0.05 P value) is presented in (9) while the graphical presentation of the significant influential factors together with 3D response surface and contour plots is given in Figures 6(a) and 6(b):

$$\text{Soil pH} = 11.07 + 0.097 * A + 1.77 * B + 0.39 * C, \tag{9}$$

where A is the polarity reversal, hr; B is the voltage gradient, V/cm; and C is the concentration, mg/kg.

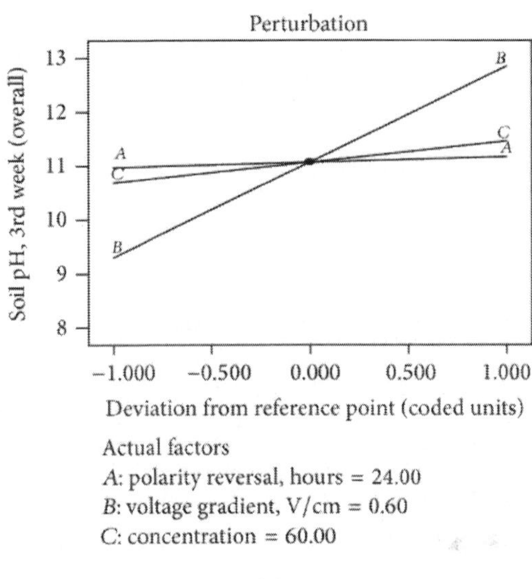

(a)

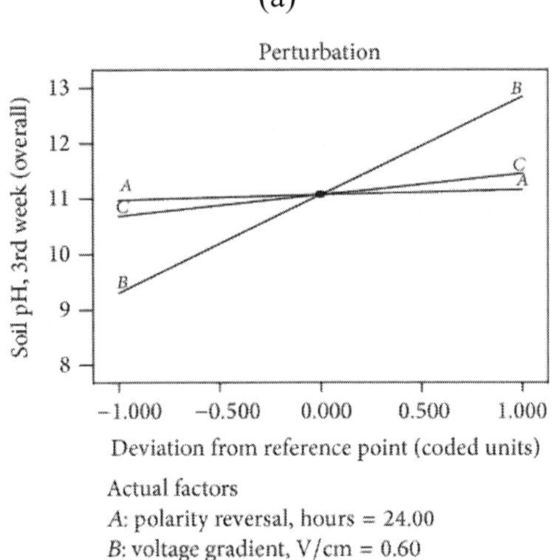

(b)

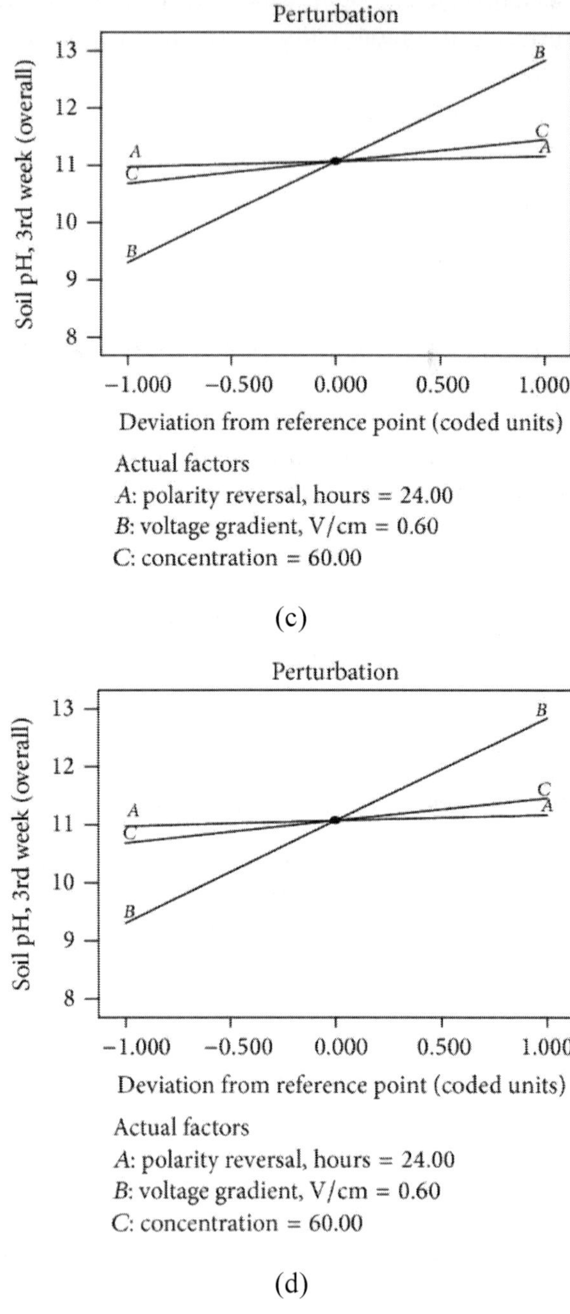

(c)

(d)

Figure 6. Perturbation plots showing the relative significance of factors on soil pH (a) and electrical conductivity (c) (left). 3D response surface and contour plots showing how the influential factors affect soil pH (b) and electrical conductivity (d) (right).

Anderson and Whitcomb [39] have reported that R^2 is biased; hence, a more accurate, less biased, and better goodness-of-fit statistic called adjusted R^2 was computed for evaluating the model accuracy. The model's R^2 and adjusted R^2 (unbiased estimate of the coefficient of determination) are 0.7725 and 0.7105, respectively. High values of R^2 are essential for modeling the experimental design space, while in identification of significant factors R^2 value does not matter and for significant factors will remain significant [39]. It is very clear that model equation, perturbation, and 3D response surface plots have shown the significant influence of voltage gradient on the soil pH over the other factors (polarity reversal rate and initial contaminant concentration). The relative contribution or effect of any given model term is directly proportional to its coefficient. Perturbation plot (Figure 6(a)) revealed a sequence of relative influence of the operating parameters on the target response as follows: voltage gradient > concentration > polarity reversal.

Bipolar Effects

The two treatment zones F and G contain 100% granular activated carbon which may be used as electrode material due to its electrical conducting properties [14]. The sides of the GAC chambers facing anode and cathode electrodes tend to behave as bipolar electrodes by acting as cathode and anode while the inner sides behave as anode and cathode, respectively.

These bipolar electrodes would be expected to generate H^+ and OH^- ions depending on whether the side is acting as anode or cathode [20] and may be expected to alter the pH distribution in the soil profile. These bipolar effects were investigated at the end of R11 and the pH profile is presented in Figure 7. The pH profile shows the variation of pH within the unspiked chambers B and D, spiked chamber C, and GAC chambers F and G. The pH ranges from 11.9 (near the anode) to 12.6 (near the cathode) which suggest that bipolar effects did not manifest due to the presence of carbonate minerals that impact high acid buffering capacity.

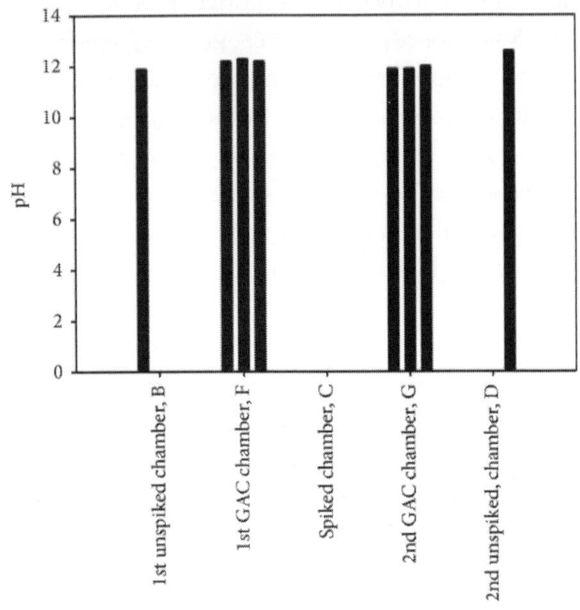

Figure 7. pH profile with two GAC treatment zones for investigating bipolar effects (R11).

Sparks [62] posited that electrical conductivity (EC) is the best index for the assessment of soil salinity. As important as this parameter is, most works on electrokinetic remediation failed to at least report the soil electrical conductivity, let alone monitor its variation over the treatment duration. Electrical conductivity greatly influences electrokinetic remediation, because it determines the amount of current flowing through the soil. The usual voltage gradient of 1 V/cm for bench-scale studies [63] when applied to saline-sodic soils would lead to high electric current flow. Lukman et al. [24] have reported that this would lead to excessive soil heating, reduction in the soil moisture content, high energy and process fluid consumption, high electroosmotic flow rate (Figure5), and in some cases higher percentage removal of contaminants. EC is simultaneously influenced by many soil properties, viz; water content, soluble salts, grain size, humus, temperature, texture, and cation exchange capacity (CEC) [64]. The 1st week of EC data shows that tests conducted using 1 V/cm (R3, 9, 6) possess the highest EC values with R1 (0.6 V/cm) coming second highest. No discernible trend was visible in the case of initial contaminant concentration despite its influence on the EC as depicted in Figure 6(c). Similar trend was observed for the 3rd week, where R9 and 6 have the highest EC values (Figure 4). A general increase of EC with time and voltage gradient (Figures6(c) and 6(d)) was observed (except for R11). The reason for this observation has been elaborately discussed above. These variations and impacts of the influential investigated factors have been modeled and presented in the 3D response surface plot in Figure 6(d). Perturbation plot (Figure6(c)) revealed a sequence of relative influence of the operating parameters on the soil electrical conductivity as follows: concentration > voltage gradient > polarity reversal.

Electroosmotic Flow

The cumulative electroosmotic volume for all the tests presented in Figure 5shows that R2 (20 mg/kg), R5 (60 mg/kg), and R4 (100 mg/kg) have the highest values. Other parameters that may influence electroosmotic flow are clay zeta potential, voltage gradient, and time-dependent fluid properties such as dielectric constant and viscosity [65]. Equation (10) shows the electroosmotic velocity as derived according to Helmholtz-Smoluchowski (H-S) theory:

$$v_e = \frac{\varepsilon_s \zeta}{\eta} E = k_e E,$$

(10)

where v_e is the electroosmotic velocity; ε_s is the pore fluid permittivity; η is the pore fluid viscosity; ζ is the soil zeta potential; k_e is the coefficient of electroosmotic conductivity; and E is the voltage gradient.

These parameters make the measured electroosmotic volume for all the tests to vary temporally. The reduction of the thickness of the diffuse double layer resulting from higher ionic concentration with subsequent higher ionic strength causes reduction in the electroosmotic flow [66]; hence higher concentrations usually yield lower electroosmotic volume (Figure 5). Reddy et al. [66] have observed similar trend. The electroosmotic volume usually decreases with time, because of the increase in electrical conductivity with time (Figure 4) that leads to higher ionic strength as the treatment proceeds. Moreover, voltage gradient has been observed to be most influential to the electroosmotic flow (Figure 8).

The least electroosmotic volumes recorded belong to the lowest voltage gradient used (0.2 V/cm), that is, in the case of R7, R12, R8, and R13. This is because high voltage gradient causes the passage of high electric current, which leads to high electromigration with subsequent substantial transfer of momentum to the surrounding pore-fluid molecules [66]. The soil zeta potential, defined as the electrical potential existing at the junction between the fixed and mobile parts of the electrical double, is influenced by the type and concentration of dissolved ions in the pore fluid in addition to the pore fluid chemistry. Clay soils, being negatively charged, usually possess negative zeta potential. At low pH below the point of zero charge (PZC), zeta potential may become positive because of excessive protonation and increase in ionic strength resulting from increased dissolution of metal ions in the pore fluid and their subsequent adsorption onto the soil particles and compression of the electrical double layer [67]. Reversal of the zeta potential charge could reverse the direction of the electroosmotic velocity as shown in (10). At high pH values, such as those encountered in this study, deprotonation and metal hydroxide precipitation could maintain a negative zeta potential; hence, electroosmotic flow will remain unidirectional as observed in all the tests. Electroosmotic flow has not been influenced by hydraulic gradient in this study as it occurs even under negative hydraulic. Equation (11) presents the model equation ($R^2 = 0.946$ and adjusted $R^2 = 0.9057$) relating the electroosmotic volume to the factors. Voltage gradient appears to be the most influential, followed by polarity reversal rate and initial contaminant concentration (Figure 8(a)). At high voltage gradient (1 V/cm), the decrease in the electroosmotic volume (Figure 8(b)) may be attributed to the development of bubbles within the electrode chambers, due to temperature rise, which then seeps into the soil to reduce the soil saturation with subsequent reduction in the electroosmotic volume [24]:

$$\text{Sqrt (Electroosmotic volume, mL)}$$
$$= 49 + 2.57 * A$$
$$+ 11.68 * B + 1.22 * C$$
$$+ 5.26 * B * C - 5.34 * A^2$$
$$- 20.95 * B^2.$$

$$(11)$$

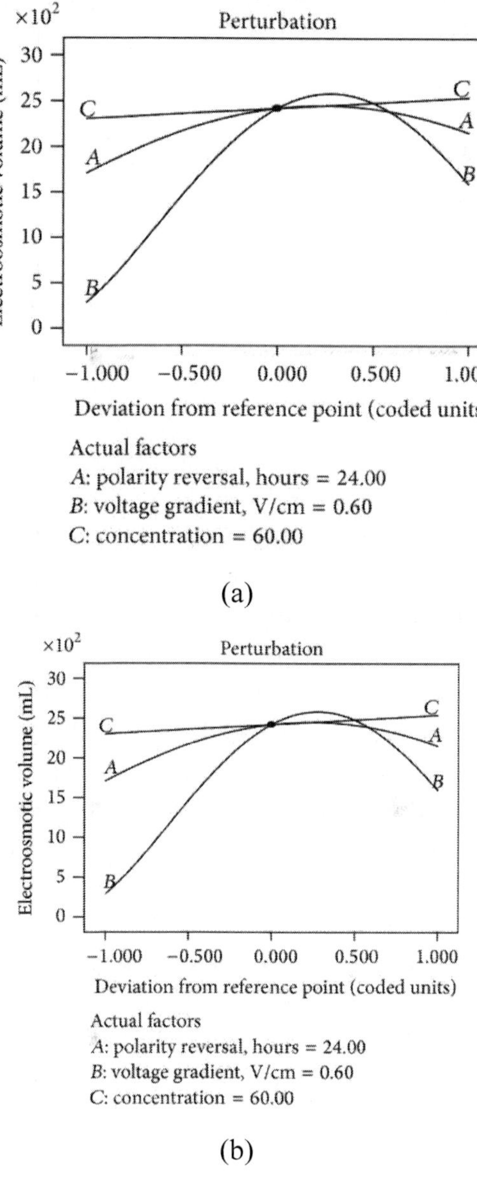

(a)

(b)

Figure 8. (a) Perturbation plot showing the relative significance of factors on electroosmotic volume. **(b)** 3D response surface and contour plots showing the influence of voltage gradient on cumulative electroosmotic volume.

Current and Temperature

Table 5 presents the average electric current recorded for each test during the 3-week test duration in descending order of magnitude to show how it is influenced by the applied voltage gradient and how it correspondingly affects the soil pH. Clearly, the higher the voltage gradient, the more amount of current is passed through soil which results in rapid generation of H^+ and OH^- ions and subsequent rise in soil pH (Table 5). The current is usually low at the beginning of the tests (Figure 9(a)), rises gradually as the tests continue, and then declines, sometimes to a stable value, while in some instances, keeps on fluctuating according to the time-dependent geochemical processes taking place such as ionic dissolution and precipitation and degradation of the processing fluids. Study conducted by Maturi and Reddy [68] corroborated the fluctuating current trend. Upon application of the driving force, the voltage gradient, the processing fluids, and pore fluid migrate while the dissolved ions electromigrate to opposite poles. These processes lead to increase in the ionic strength of the pore fluid thereby increasing the current flow to a maximum value. The observed decline of the current to a stable value may be attributed to the electromigration of cations and anions to the respective electrode with subsequent possible precipitation of the cations due to increase in the soil pH as the test progresses [66, 69]. Temporal geochemical processes such as mineral and chemical dissolution and neutralization reactions taking place in the electrode chambers also contribute to the variation of the electric current. A maximum value of 5.13 A was recorded for R6 whose average current was 3.02 A. This current is considered extremely high, considering the fact that it is about two orders of magnitude greater than the recorded current values for other bench-scale studies that employed the Lasagna process (<30 mA) in other soil apart from saline-sodic soil as shown in Table 1. Other studies using electrokinetic remediation only using voltage gradient of 1 V/cm or higher have reported higher values but usually less than 300 mA [58, 66, 70]. Using low voltage gradient of 0.2 V/cm has only resulted in reducing the current to about 130–210 mA (Table 5). This unique and important observation may be explained by the high salinity and sodicity of the investigated soil which provides large amount of dissolved salts and minerals (carbonates) in the pore fluid for sustained high electrical conduction. High current flow through the soil will significantly affect the soil temperature, electroosmotic flow rate, electrode material and processing fluids degradation, soil pH, geochemical processes, remedial efficiency and energy consumption. In a related study by Lukman et al. [24], they also recorded similar high current (2.8 A). To emphasize on the effect of the electric current on the soil temperature, current and temperature readings recorded using a time step of 30 min is presented in Figure 9 for R11 (voltage gradient = 0.6 V/cm). This test has 0.61 A and 28.45°C as the average current and temperature respectively. The maximum values were 0.91 A and 34.6°C respectively which were recorded under room temperature of 24°C. It is clear from Figure 9 that low current leads to low soil temperature and vice-versa. In a preliminary study conducted by Lukman et al. [24] using 1 V/cm, 36.34°C, and 47°C were the average and maximum soil temperatures, indicating that the soil becomes very hot when using 1 V/cm. While soil heating may be advantageous in increasing the volatility of organics, solubility of minerals (carbonates), and reduction in pore fluid viscosity which will increase electroosmotic flow, it may also be undesirable since it will reduce the soil moisture content due to pore fluid evaporation with subsequent reduction in current and electroosmotic flow. In addition, it will increase soil electrical conductivity and energy expenditure [20]. Previous studies have not reported significant rise of soil temperature during bench-scale tests [20]. A linear model was obtained (12) which relates the factors to the average electric current whose respective R^2 and adjusted R^2 are 0.9556 and 0.9435. The perturbation and response surface plots (Figure 10) also revealed the significant influence of the applied voltage gradient over initial contaminant concentration and polarity reversal rate:

$$\text{Sqrt (Average current)} = 1 + 0.020 * A + 0.59 * B - 0.059 * C.$$

$$(12)$$

Table 5. Comparing electrical current with voltage gradient and soil pH for all tests.

Run	Current, A	Voltage gradient, V/cm	pH
R6	3.02	1	12.9
R9	2.65	1	12.6
R3	2.25	1	12.6
R4	2.04	1	12.7
R2	1.32	0.6	10.9
R1	1.17	0.6	10.2
R10	1.12	0.6	11.9
R5	1.03	0.6	11.2
R11	0.61	0.6	12
R8	0.21	0.2	9.8
R12	0.15	0.2	8.3
R7	0.14	0.2	8.1
R13	0.13	0.2	10.4

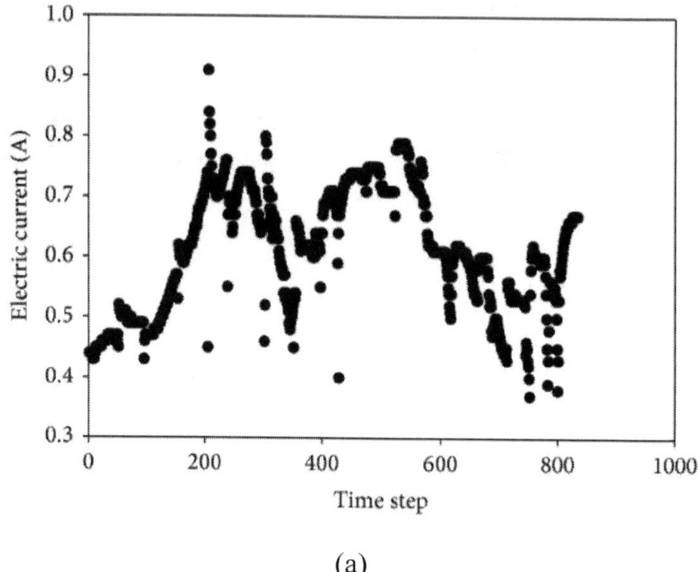

(a)

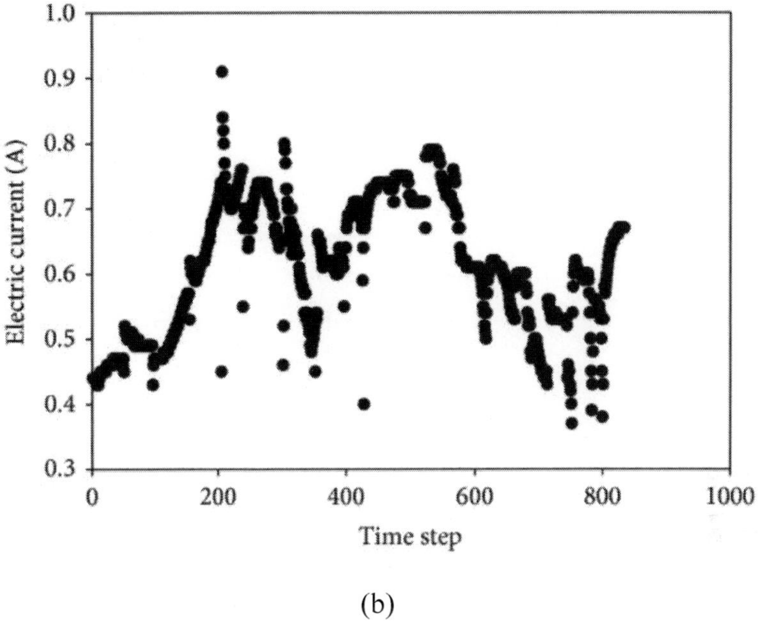

(b)

Figure 9. Comparing variations of electric current with soil temperature: (a) current; (b) temperature.

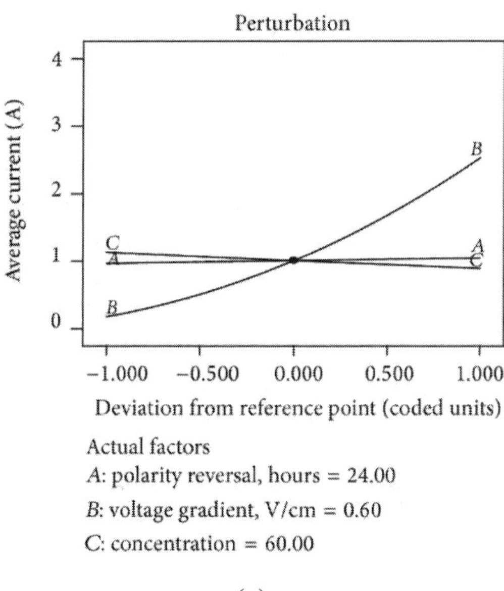

(a)

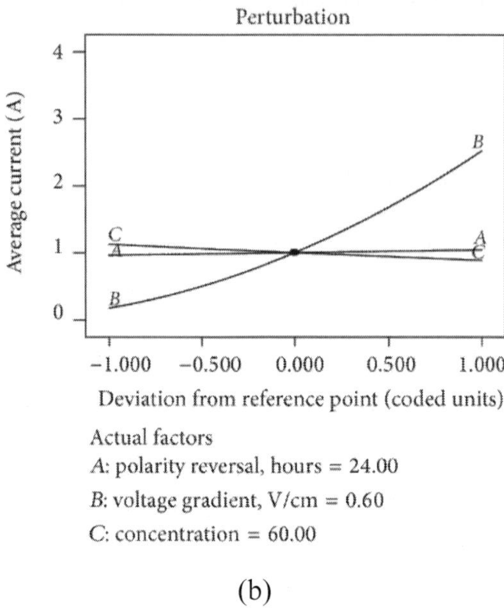

(b)

Figure 10. (a) Perturbation plot showing the relative significance of factors on average electric current. (b) 3D response surface and contour plots showing the influence of voltage gradient on average electric current.

Trivalent Chromium Migration, Model Validation, and Optimization

Figure 11 presented the distribution and migration of trivalent Cr from the contaminated chamber, C, to the GAC chambers F and G for all the thirteen (13) tests. This migration becomes more pronounced for tests R5, R6, and R9. In the case of R6 (no polarity reversal), significant trivalent Cr migration took place from the contaminated chamber, C, to the GAC chamber, F, near the anode. This observation may be attributed to the formation of high amount of negatively charged metal hydroxocomplexes at pH 12.9, which are then attracted to the anode via electromigration but become adsorbed onto the GAC in chamber F during the transport process. Visual MINTEQ 3.0 [51] was employed to model the trivalent Cr ion speciation for R5 from the weekly monitoring data using the dissolved concentration, pH, temperature and ionic strength. The speciation diagram presented in Figure 12 reveals the increasing dominance of the negatively charged complex $Cr(OH)_4^-$ and the decreasing concentration of aqueous

$Cr(OH)_3$ at pH 11.2. This explains the greater movement of the trivalent Cr species toward the anode in R6 at pH 12.9. Pourbaix [71] and Chinthamreddy and Reddy [29] have already asserted that $Cr(OH)_4^-$ ions will become the dominant species at pH values greater than 11.8, thus, trivalent Cr solubility increases. However, under normal soil pH, trivalent Cr has limited solubility and highly adsorbs to soil [29, 32]. In a related study by Reddy and Chinthamreddy [30] which involved an alkaline and high acid buffering soil called glacial till, they did not observe significant trivalent Cr migration and no removal. Although, the soil redox state may be dynamic because of the generation of oxygen and hydrogen gases at the electrodes in addition to the possible presence of iron (reducing agent), manganese (oxidizing agent) or microorganisms that can oxidize the trivalent Cr to the hexavalent form; oxidation of trivalent Cr does not take place appreciably in high buffering capacity soil such as saline-sodic soil [28]. For this reason, hexavalent Cr was not studied. Migration of the trivalent Cr from the contaminated chamber to the GAC chambers indicated remarkable remedial efficiency for some of the tests (R5, R6 and R9) while others indicated low or no removal at all (R1–R4, R7, R10, and R12). There is zero remedial efficiency when there was accumulation of the contaminant at the sampling location thereby having the residual concentration (C_o) to be greater than the initial (C), in which case, $C_o/C > 1$. Hence Figure 11 utilized C_o/C to indicate the migration of trivalent Cr when $C_o/C < 1$ or its accumulation at any given location or chamber when $C_o/C > 1$.

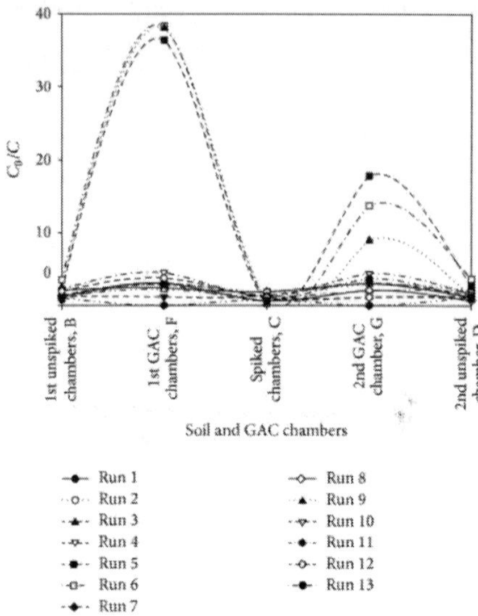

Figure 11. Trivalent Cr distribution and migration from the contaminated chamber to the GAC chambers after 13 tests.

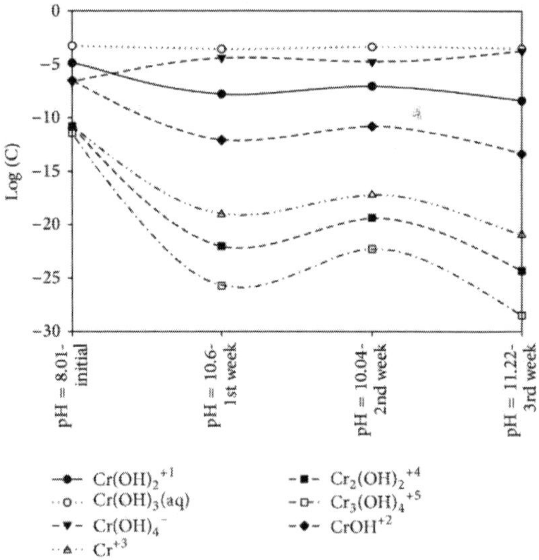

Figure 12. Speciation diagram for trivalent Cr species at different weekly pH values.

Mass balance analyses of Cr were performed for Runs 8, 11, and 13. From Table 6, the mass balance for Runs 8, 11, and 13 is 121.75, 74.51, and 148.99%, respectively. These values were obtained using the ratio between the residual Cr in the contaminated chamber (C) plus any increase in Cr concentration in the GAC chamber and the initial Cr concentration. Among other reasons for the discrepancies in mass balance that is sometimes encountered during electrokinetic remediation as put forward by previous investigators [30, 66, 72] include adsorption onto the electrode and geotextile materials (which houses the GAC in the two chambers) and non-uniform distribution of contaminants within the small soil sample (about 2 g) taken for acid digestion and analysis. Taking different samples spatially from the contaminated chamber will help improve the mass balance.

Table 6. A sample mass balance analysis of trivalent Cr for Runs 8, 11, and 13

Runs	Run 8	Run 11	Run 13
Initial concentration, mg/kg	37.2	77.95	37.2
Residual concentration, mg/kg	23.46	58.08	24.23
1st GAC chamber, F			
Initial concentration, mg/kg	6.9	6.9	6.9
Residual concentration, mg/kg	21.15	0	19.7
2nd GAC chamber, G			
Initial concentration, mg/kg	6.9	6.9	6.9
Residual concentration, mg/kg	14.48	0	25.3
Mass balance, %	121.75	74.51	148.99

The tests were sorted in decreasing order of remedial efficiency (Table 7) to reveal some salient points that will help in providing adequate connection between factors and responses. Highest remedial efficiencies (79.97–34.88%) were recorded for tests involving 60 mg/kg initial trivalent Cr concentration, whereas no removal was recorded for all tests involving 20 mg/kg. Only one test involving 100 mg/kg recorded some remedial efficiency (Table 7). Low remedial efficiency at 20 mg/kg may be attributed to the availability of adsorption sites for trivalent Cr ions coupled with the high selectivity for Cr for this particular soil type [45] at the given concentration. At higher concentrations (100 mg/kg) and pH, trivalent Cr may precipitate as , thus, rendering it immobile [66]. Even with low electric current, electroosmotic flow and voltage gradient (0.2 V/cm), 34.88% and 36.93% of the trivalent Cr was removed from the contaminated chamber in tests R13 and R8, respectively. Polarity reversal rate did not show any discernible pattern. Hence, there is need for simultaneous optimization of these three factors for optimal removal of the trivalent Cr. It is important to note that high voltage gradient (1 V/cm) or passage of high electric current does not necessarily translate into high remedial efficiency but will definitely increase the energy expenditure. At high voltage gradient, current is high, leading to high electroosmotic flow toward cathode. This opposite flow may interfere with the electromigration of the anionic trivalent Cr species that are migrating toward the anode, thus, reducing the overall remedial efficiency. Electromigration constitute the major transport mechanism for charged species whose rate is 10–300 times higher than the advective electroosmotic transport [73]. At low voltage gradient (0.2 V/cm), extremely low electroosmotic flow takes place and sustained electromigration prevails. The weekly percentage removal of trivalent Cr from the contaminated chamber is presented in Figure 13. The dynamic and temporal changes in the geochemical processes controlling the contaminant removal are attributable to the observed trends in the weekly percentage removal.

Table 7. Comparing trivalent Cr remedial efficiency with factors and some responses

Runs	Remedial efficiency, %	Current, A	Residual, pH	Electroosmotic volume, mL	Polarity reversal rate, hr	Voltage gradient, V/cm	Initial Cr concentration, mg/kg
R5	79.97	1.03	11.2	2344.5	24	0.6	60
R6	72.73	3.02	12.9	1201.5	0	1	60
R9	65.66	2.65	12.6	1399.5	48	1	60
R8	36.93	0.21	9.8	81	0	0.2	60
R13	34.88	0.13	10.4	63	48	0.2	60
R11	25.5	0.61	12	1387.84	0	0.6	100
R1	0	1.17	10.2	1728	0	0.6	20
R2	0	1.32	10.9	2542.5	48	0.6	20
R3	0	2.25	12.6	814.5	24	1	20
R4	0	2.04	12.7	2272.5	24	1	100
R7	0	0.14	8.1	396	24	0.2	20
R10	0	1.12	11.9	2236.5	48	0.6	100
R12	0	0.15	8.3	324	24	0.2	100

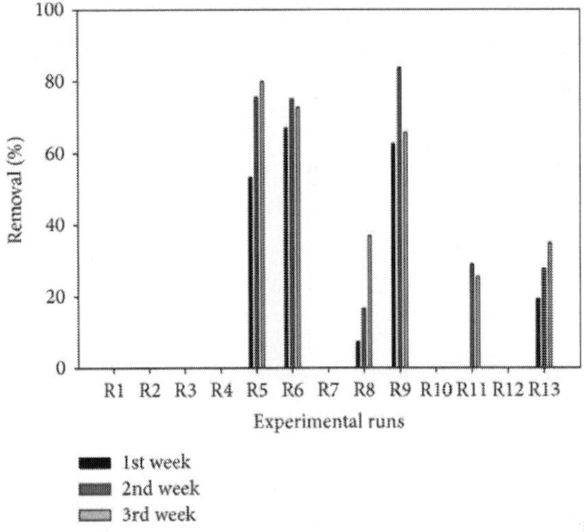

Figure 13. Weekly percentage removal of trivalent Cr for 13 tests.

Equation (13) relates the investigated factors to the remedial efficiency with 0.9335 and 0.8966 as the R^2 and adjusted R^2 values, respectively:

$$\text{Sqrt (Cr, remedial efficiency)}$$
$$= 8.78 - 0.71 * A + 0.58 * B$$
$$+ 0.63 * C - 1.50 * B^2 - 7.39 * C^2.$$

$$(13)$$

Perturbation plot (Figure 14(a)) also supports the observed influence of the initial Cr concentration on the remedial efficiency, followed by voltage gradient, then, polarity reversal rate. The investigated factor levels can be used to determine the optimal conditions required to achieve maximum remedial efficiency as depicted in the 3D response surface plot (Figure 14(b)).

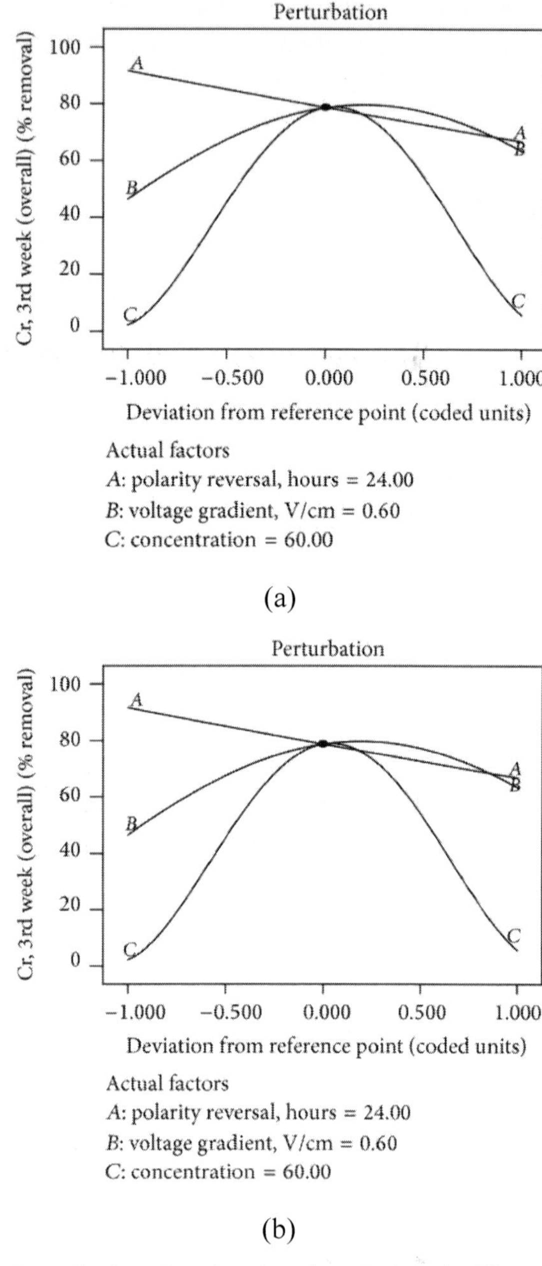

Figure 14. (a) Perturbation plot showing the relative significance of factors on trivalent Cr remedial efficiency. (b) 3D response surface and contour plots showing the influence of initial contaminant concentration on trivalent Cr remedial efficiency.

Model Validation

To validate the practical applicability of the developed models affecting the remedial efficiency (13) and soil pH (9), additional experimental test was run at voltage gradient of 1 V/cm, initial contaminant concentration of 44.15 mg/kg, and without polarity reversal (Table 8). Results of the model validation showed that the experimental results lie within 90% confidence interval (CI) and prediction interval (PI) with associated prediction error of 2.35% and 32.64% for soil pH and remedial efficiency, respectively. Since the validation results fall within the prediction interval, then, the outcome of the confirmation test was a success [39]. Hence, the models can provide good approximations necessary to move in the proper direction.

Table 8. Experimental validation of trivalent Cr remedial efficiency and soil pH using voltage gradient = 1 V/cm; average concentration = 44.15 mg/kg; and polarity reversal rate = 0 hr.

Response	Experimental result	Model prediction	Prediction error, %	90% CI* low	90% CI high	90% PI** low	90% PI high
Cr, remedial efficiency	75.88	51.11	32.64	31.17	75.95	18.36	100
Residual soil, pH	12.3	12.6	2.35	11.7	13.5	10.8	14

Confidence interval.

**Prediction interval.

Optimization of Trivalent Chromium Removal

Numerical optimization was employed to find the optimal factor levels that will specifically target maximum remedial efficiency of trivalent Cr while optimizing all the other contaminant remedial efficiencies and responses (Figure 15). An overall desirability value of 0.715 was obtained and its variation based on the influential factors (initial concentration and voltage gradient) is depicted in Figure 16. Optimal conditions required to achieve effective trivalent Cr removal at 60 mg/kg are presented in Table 9. Overall desirability of 0.715 was attained at the following optimal conditions: voltage gradient = 0.36 V/cm; polarity reversal rate = 17.63 hr;

soil pH = 10.0. Under these conditions, the expected trivalent Cr remedial efficiency is 64.75%.

Table 9. Optimal factor levels required to maximize remedial efficiency of trivalent Cr.

Item	Value
Polarity reversal, hours	17.63
Voltage gradient, V/cm	0.36
Concentration, mg/kg	60
Expected remedial efficiency of trivalent Cr	64.75
Expected residual soil pH	10
Desirability	0.715

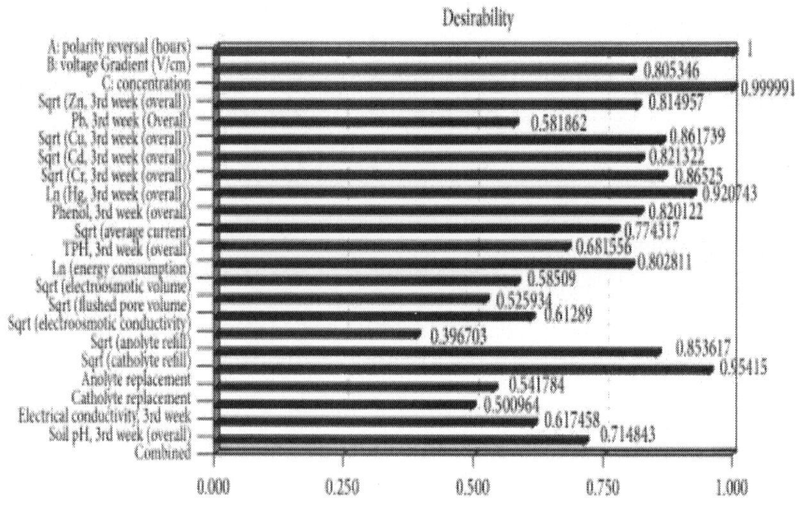

Figure 15. Combined and individual response desirability values for all responses and factors.

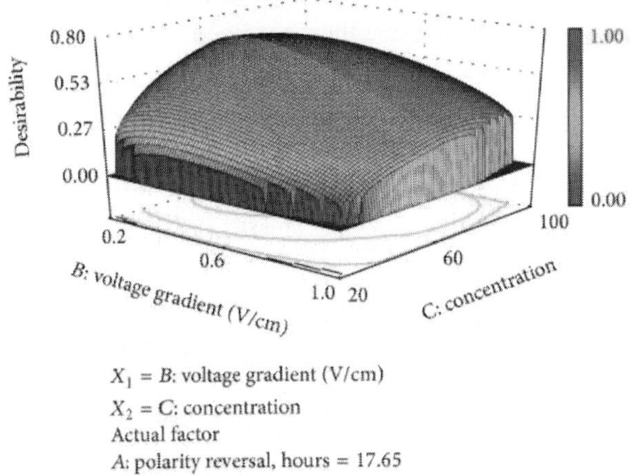

X_1 = B: voltage gradient (V/cm)
X_2 = C: concentration
Actual factor
A: polarity reversal, hours = 17.65

Figure 16. 3D surface plot of the overall desirability variation relative to influential factors.

Impacts of the Integrated Electrokinetic Remediation on Soil Physicochemical Properties

Preceding sections have elaborately discussed and modeled the impacts of the proposed remediation technique on the soil pH and electrical conductivity. Additionally, the passage of electric current and soil pH gradients will result in the following physicochemical interactions: (1) possible dissolution of the clay minerals beyond a pH range of 7–9; (2) dissolution of available soil salts such as carbonates; (3) production of cementitious products resulting from the precipitation of metal ions at pH values corresponding to their hydroxide solubility values; and (4) soil structural changes which affect its engineering characteristics [41–44]. Surface area, pore volume and size (Table 10), mineralogical compositions (Table 11), and elemental constituents (Table 12) were analyzed, before and after the test for R5. At the end of the test (pH = 11.2), the soil specific surface area has increased (9.07 to 11.21 m^2/g) with corresponding increase in the pore volume and size. These results have confirmed that some dissolution of the soil minerals has taken place during the electrokinetic remediation process due to variations in the pore fluid chemistry. Soil pores are due to the presence of interlayer spaces that becomes prominent in 2 : 1 clay mineral types such as montmorillonite and smectite [40, 62, 74]. Table 11 presents the mineral transformation where dolomite completely disappeared; calcite and quartz were altered and degraded, respectively, after the test. The constituent soil elements were not spared as the amount of each one either increased or decreased after the test as shown in Table 12. These observations may be explained by microbially-driven biotransformation processes involving dissolution and precipitation, which take place under both aerobic anaerobic conditions. This leads to mineral dissolution and formation of new minerals from aqueous ions (biomineralization) as noticed in Table 11 [40]. Yong et al. [40] have asserted that the scientific basis for biomineralization is still not well understood.

Table 10. Values of soil surface area and pore volume and size, before and after treatment.

Description	BET* surface area, m2/g	Pore volume, cm3/g	Pore size, Å
Before	9.07	0.014	62.55
After	11.21	0.045	163.24

*BET: Brunauer-Emmett-Teller.

Table 11. Soil mineralogical transformations before and after treatment

Phase name	Before, %	After, %
Quartz, SiO_2	87.4	55.3
Calcite, $CaCO_3$	5.2	44.7
Dolomite, $CaMg(CO_3)_2$	7.4	—

Table 12. Values of constituent soil elements, before and after treatment

Element	Before, %	After, %
Ca	37.64	42.06
Si	34.73	23.42
Fe	10.41	15.06
Al	7.6	9.55
K	3.42	4.61
Mg	2.48	2.49
Pd	2.85	1.46
Ti	0.86	1.35

CONCLUSIONS

The study reported herein investigated the migration of trivalent Cr ions from a multiple contaminated natural saline-sodic soil. The soil salinity and sodicity, which provided large amount of dissolved salts and minerals (carbonates) in the pore fluid for sustained high electrical conduction, were responsible for the extremely high electric current flow. This led to excessive soil heating, high energy and process fluid consumption, high electroosmotic volume, and in some cases higher percentage removal of trivalent Cr. Significant migration of Cr from the contaminated chamber to the granular activated carbon chamber was recorded which led to highest remedial efficiencies (79.97–34.88%) for tests involving 60 mg/kg initial trivalent Cr concentration, whereas no removal was recorded for all tests involving 20 mg/kg. Even under low electric current, electroosmotic flow, and voltage gradient (0.2 V/cm), up to 36.93% of the trivalent Cr was removed from the contaminated chamber. It has been shown that high voltage gradient (1 V/cm) or passage of high electric current does not necessarily translate into high remedial efficiency. Bipolar effects did not manifest due to the presence of carbonate minerals that impact high acid buffering capacity. For test without polarity reversal, trivalent Cr moved toward the anode due to the formation of high amount of anionic $Cr(OH)_4^-$ hydroxocomplex at high pH, which was further attracted to the anode via electromigration. Nonadsorption of this ion onto the negatively charged clay soil due to the possession of similar charge increased its availability and mobility. Speciation modeling using Visual MINTEQ 3.0 reveals the increasing dominance of the anionic $Cr(OH)_4^{--}$ and the decreasing concentration of aqueous $Cr(OH)_3$ at pH 11.2. Effects of voltage gradient, initial contaminant concentration, and polarity reversal rate on the effective removal of Cr ions were experimentally studied using the Box-Behnken Design of experiment and mathematically modeled and numerically optimized using response surface methodology. Results of the model validation showed that the experimental results lie within 90% confidence interval and prediction interval with associated prediction error of 2.35% and 32.64% for soil pH and trivalent Cr remedial efficiency, respectively. Overall desirability of 0.715 was attained at the following optimal conditions: voltage gradient = 0.36 V/cm; polarity reversal rate =

17.63 hr; and soil pH = 10.0. Under these conditions, the expected trivalent Cr remedial efficiency is 64.75%. Passage of electric current and variations in the pore fluid chemistry led to soil mineral dissolution and alteration via biotransformation.

ACKNOWLEDGMENTS

The authors would like to acknowledge the support provided by King Abdul-Aziz City for Science and Technology (KACST) through the Science & Technology Unit at King Fahd University of Petroleum & Minerals (KFUPM) for funding this work through Project no. 11-Env1669-04, as part of the National Science, Technology and Innovation Plan.

REFERENCES

1.   A. Alok, R. P. Tiwari, and R. P. Singh, "Effect of pH of anolyte in electrokinetic remediation of cadmium contaminated soil," International Journal of Engineering Research & Technology, vol. 1, pp. 1–11, 2012.
2.   A. N. Puri and B. Anand, "Reclamation of alkali soils by electrodialysis," Soil Science, vol. 42, no. 1, pp. 23–27, 1936.
3.   A. T. Yeung and Y. Gu, "A review on techniques to enhance electrochemical remediation of contaminated soils," Journal of Hazardous Materials, vol. 195, pp. 11–29, 2011.
4.   A. T. Yeung, "Geochemical processes affecting electrochemical remediation," inElectrochemical Remediation Technologies for Polluted Soils, Sediments and Groundwater, pp. 65–94, John Wiley & Sons, 2009.
5.   A. T. Yeung, "Milestone developments, myths, and future directions of electrokinetic remediation," Separation and Purification Technology, vol. 79, no. 2, pp. 124–132, 2011.
6.   A. Z. Al-Hamdan and K. R. Reddy, "Transient behavior of heavy metals in soils during electrokinetic remediation," Chemosphere, vol. 71, no. 5, pp. 860–871, 2008.
7.   B. D. Swift and J. J. Tarantino, "Application of the Lasagna soil remediation technology at the DOE paducah gaseous diffusion plant," in Proceedings of the Waste Management Symposium (WM '03), Tucson, Ariz, USA, February 2003.

8. C. D. Palmer and P. R. Wittbrodt, "Processes affecting the remediation of chromium-contaminated sites," Environmental Health Perspectives, vol. 92, pp. 25–40, 1991.

9. C. J. Athmer and S. V. Ho, "Field studies: organic-contaminated soil remediation with lasagna technology," in Electrochemical Remediation Technologies for Polluted Soils, Sediments and Groundwater, K. R. Reddy and C. Cameselle, Eds., pp. 625–646, John Wiley & Sons, 2009.

10. D. B. Gent, R. M. Bricka, A. N. Alshawabkeh, S. L. Larson, G. Fabian, and S. Granade, "Bench- and field-scale evaluation of chromium and cadmium extraction by electrokinetics,"Journal of Hazardous Materials, vol. 110, no. 1–3, pp. 53–62, 2004.

11. D. Derringer and R. Suich, "Simultaneous optimization of several response variables,"Journal of Quality Technology, vol. 12, pp. 214–219, 1980.

12. D. L. Sparks, Environmental Soil Chemistry, Academic Press, New York, NY, USA, 2nd edition, 2003.

13. EPA, Method 3050B—Acid Digestion of Sediments, Sludges, and Soils, United States Environmental Protection Agency, 1996.

14. F. E. Asimakopoulou, I. F. Gonos, and I. A. Stathopulos, "Methodologies for determination of soil ionization gradient," Journal of Electrostatics, vol. 70, no. 5, pp. 457–461, 2012.

15. F. F. Reuss, "Sur un nouvel effet de l'électricité galvanique," Mémoires de la Societé Impériale des Naturalistes de Moscou, vol. 2, pp. 327–337, 1809.

16. H. A. Abramson, Electrokinetic Phenomena and Their Application to Biology and Medicine, ACS Monograph Series, Chemical Catalog, New York, NY, USA, 1934.

17. I. P. Abrol, J. S. P. Yadav, and F. I. Massoud, "Salt-affected soils and their management, food and agriculture Organization of the United Nations," FAO Soils Bulletin, vol. 39, 1988.

18. I. Ravina and D. Zaslavsky, "Non-linear electrokinetic phenomena—I: review of literature,"Soil Science, vol. 106, pp. 60–66, 1968.

19. J. Hamed, Y. B. Acar, and R. J. Gale, "Pb(II) removal from kaolinite by electrokinetics,"Journal of geotechnical engineering, vol. 117, no. 2, pp. 241–271, 1991.

20. J. K. Mitchell, Fundamentals of Soil Behavior, John Wiley & Sons, 1993.

21. J. M. Dzenitis, "Steady state and limiting current in electroremediation of soil," Journal of the Electrochemical Society, vol. 144, no. 4, pp. 1317–1322, 1997.

22. J. P. Gustafsso, "Visual MINTEQ ver. 3.0.," 2010,http://www2.lwr.kth.se/English/OurSoftware/vminteq/index.htm.

23. J. Virkutyte, M. Sillanpää, and P. Latostenmaa, "Electrokinetic soil remediation—critical overview," Science of the Total Environment, vol. 289, no. 1–3, pp. 97–121, 2002.

24. J. W. Ma, F. Y. Wang, Z. H. Huang, and H. Wang, "Simultaneous removal of 2,4-dichlorophenol and Cd from soils by electrokinetic remediation combined with activated bamboo charcoal," Journal of Hazardous Materials, vol. 176, no. 1–3, pp. 715–720, 2010.

25. J. W. Ma, H. Wang, and Q. Luo, "Movement-adsorption and its mechanism of Cd in soil under combining effects of electrokinetics and a new type of bamboo charcoal," Chinese Journal of Environmental Science, vol. 28, no. 8, pp. 1829–1834, 2007 (Chinese).

26. J. W. Ma, H. Wang, and R. R. Li, "Removal of cadmium in kaolin by electrokinetics-bamboo charcoal adsorption," Environmental Chemistry, vol. 26, pp. 634–637, 2007 (Chinese).

27. K. Beddiar, T. Fen-Chong, A. Dupas, Y. Berthaud, and P. Dangla, "Role of pH in electro-osmosis: experimental study on NaCl-water saturated kaolinite," Transport in Porous Media, vol. 61, no. 1, pp. 93–107, 2005.

28. K. K. Reddy and S. Chinthamreddy, "Effects of initial form of chromium on electrokinetic remediation in clays," Advances in Environmental Research, vol. 7, no. 2, pp. 353–365, 2003.

29. K. Maturi and K. R. Reddy, "Cosolvent-enhanced desorption and transport of heavy metals and organic contaminants in soils during electrokinetic remediation," Water, Air, and Soil Pollution, vol. 189, no. 1–4, pp. 199–211, 2008.

30. K. Maturi and K. R. Reddy, "Simultaneous removal of organic compounds and heavy metals from soils by electrokinetic remediation with a modified cyclodextrin," Chemosphere, vol. 63, no. 6, pp. 1022–1031, 2006.

31. K. R. Reddy and M. R. Karri, "Effect of voltage gradient on integrated electrochemical remediation of contaminant mixtures," Land Contamination and Reclamation, vol. 14, no. 3, pp. 685–698, 2006.

32. K. R. Reddy and S. Chinthamreddy, "Enhanced electrokinetic remediation of heavy metals in glacial till soils using different electrolyte solutions," Journal of Environmental Engineering, vol. 130, no. 4, pp. 442–455, 2004.

33. K. R. Reddy and U. S. Parupudi, "Removal of chromium, nickel and cadmium from clays by in-situ electrokinetic remediation," Soil and Sediment Contamination, vol. 6, no. 4, pp. 391–407, 1997.

34. K. R. Reddy, P. R. Ala, S. Sharma, and S. N. Kumar, "Enhanced electrokinetic remediation of contaminated manufactured gas plant soil," Engineering Geology, vol. 85, no. 1–2, pp. 132–146, 2006.

35. K. R. Reddy, R. E. Saichek, K. Maturi, and P. Ala, "Effects of soil moisture and heavy metal concentrations on electrokinetic remediation," Indian Geotechnical Journal, vol. 32, pp. 258–288, 2002.

36. K. R. Reddy, S. Chinthamreddy, R. E. Saichek, and T. J. Cutright, "Nutrient amendment for the bioremediation of a chromium-contaminated soil by electrokinetics," Energy Sources, vol. 25, no. 9, pp. 931–943, 2003.

37. K. R. Reddy, S. Danda, and R. E. Saichek, "Complicating factors of using ethylenediamine tetraacetic acid to enhance electrokinetic remediation of multiple heavy metals in clayey soils," Journal of Environmental Engineering, vol. 130, no. 11, pp. 1357–1366, 2004.

38. K. R. Reddy, U. S. Parupudi, S. N. Devulapalli, and C. Y. Xu, "Effects of soil composition on the removal of chromium by electrokinetics," Journal of Hazardous Materials, vol. 55, no. 1–3, pp. 135–158, 1997.

39. L. Hopkinson, A. Cundy, D. Faulkner, A. Hansen, and R. Pollock, "Electrokinetic stabilization of chromium (VI)-contaminated soils, electrochemical remediation technologies for polluted soils," in Sediments and Groundwater, pp. 179–193, 2009.

40. L. M. Vane and G. M. Zang, "Effect of aqueous phase properties on clay particle zeta potential and electro-osmotic permeability: implications for electro-kinetic soil remediation processes," Journal of Hazardous Materials, vol. 55, no. 1–3, pp. 1–22, 1997.

41. M. A. Bezerra, R. E. Santelli, E. P. Oliveira, L. S. Villar, and L. A. Escaleira, "Response surface methodology (RSM) as a tool for optimization in analytical chemistry," Talanta, vol. 76, no. 5, pp. 965–977, 2008.

42. M. Elektorowicz, "Electrokinetic remediation of mixed metals and organic contaminants," inElectrochemical Remediation Technologies for Polluted Soils, Sediments and Groundwater, pp. 315–331, John Wiley & Sons, 2009.

43. M. H. Essa and M. A. Al-Zahrani, "Date pits as potential raw materials for the production of activated carbons in Saudi Arabia," International Journal of Applied Environmental Sciences, vol. 4, no. 1, pp. 47–58, 2009.

44. M. H. Essa, M. A. Al-Zahrani, and T. N. Suresh, "Optimization of activated carbon production from date pits," International Journal of Environmental Engineering, vol. 5, pp. 325–338, 2013.

45. M. J. Anderson and P. J. Whitcomb, RSM Simplified: Opitimizing Processes Using Response Surface Methods for Design of Experiments, Productivity Press, 2005.

46. M. Pazos, A. Plaza, M. Martín, and M. C. Lobo, "The impact of electrokinetic treatment on a loamy-sand soil properties," Chemical Engineering Journal, vol. 183, pp. 231–237, 2012.

47. M. Pourbaix, Atlas of Electrochemical Equilibria in Aqueous Solutions, 1974.

48. P. H. Brodsky and S. V. Ho, "In situ remediation of contaminated soils," U.S. Patent 5.398.756, 1995.

49. P. R. Buchireddy, R. M. Bricka, and D. B. Gent, "Electrokinetic remediation of wood preservative contaminated soil containing copper, chromium, and arsenic," Journal of Hazardous Materials, vol. 162, no. 1, pp. 490–497, 2009.

50. Q.-Y. Wang, D.-M. Zhou, L. Cang, and T.-R. Sun, "Application of bioassays to evaluate a copper contaminated soil before and after a pilot-scale electrokinetic remediation,"Environmental Pollution, vol. 157, no. 2, pp. 410–416, 2009.

51. R. H. Myers, D. C. Montgomery, and C. M. Anderson-Cook, Response Surface Methodology: Process and Product Optimization Using Designed Experiments, John Wiley & Sons, 2009.

52. R. J. Hunter and M. James, "Charge reversal of kaolinite by hydrolyzable metal ions: an electroacoustic study," Clays & Clay Minerals, vol. 40, no. 6, pp. 644–649, 1992.

53. R. N. Yong, M. Nakano, and R. Pusch, Environmental Soil Properties and Behaviour, CRC Press, New York, NY, USA, 2012.

54. S. Chinthamreddy and K. R. Reddy, "Geochemistry of chromium during electrokinetic remediation," in Proceedings of the 4th International Symposium on Environmental Geotechnology and Global Sustainable Development, Boston, Mass, USA, 1998.

55. S. Chinthamreddy and K. R. Reddy, "Oxidation and mobility of trivalent chromium in manganese-enriched clays during electrokinetic remediation," Soil and Sediment Contamination, vol. 8, no. 2, pp. 197–216, 1999.

56. S. H. Kim, H. Y. Han, Y. J. Lee, C. W. Kim, and J. W. Yang, "Effect of electrokinetic remediation on indigenous microbial activity and community within diesel contaminated soil," Science of the Total Environment, vol. 408, no. 16, pp. 3162–3168, 2010.

57. S. Lukman, M. H. Essa, N. D. Mu'azu, A. Bukhari, and C. Basheer, "Adsorption and desorption of heavy metals onto natural clay material: influence of initial pH," Journal of Environmental Science and Technology, vol. 6, no. 1, pp. 1–15, 2013.

58. S. Lukman, M. H. Essa, N. D. Mu'azu, and A. Bukhari, "Coupled electrokinetics-adsorption technique for simultaneous removal of heavy metals and organics from saline-sodic soil," The Scientific World Journal, vol. 2013, Article ID 346910, 9 pages, 2013.

59. S. Pamukcu, "Electrochemical transport and transformations," in Electrochemical Remediation Technologies for Polluted Soils, Sediments and Groundwater, pp. 29–65, John Wiley & Sons, New York, NY, USA, 2009.

60. S. V. Ho and P. H. Brodsky, "In-situ remediation of contaminated heterogeneous soils," U.S. Patent 5,476,992, 1995.

61. S. V. Ho, B. M. Hughes, P. H. Brodsky, J. S. Merz, and L. P. Egley, "Advancing the use of an innovative cleanup technology: case study of Lasagna," Remediation Journal, vol. 9, pp. 103–116, 1999.

62. S. V. Ho, C. Athmer, P. W. Sheridan et al., "The lasagna technology for in situ soil remediation. 1. Small field test," Environmental Science and Technology, vol. 33, no. 7, pp. 1086–1091, 1999.

63. S. V. Ho, C. Athmer, P. W. Sheridan et al., "The lasagna technology for in situ soil remediation. 2. Large field test," Environmental Science & Technology, vol. 33, no. 7, pp. 1092–1099, 1999.

64. S. V. Ho, C. J. Athmer, P. W. Sheridan, and A. P. Shapiro, "Scale-up aspects of the Lasagna process for in situ soil decontamination," Journal of Hazardous Materials, vol. 55, no. 1–3, pp. 39–60, 1997.

65. S. V. Ho, P. W. Sheridan, C. J. Athmer et al., "Integrated in situ soil remediation technology: the Lasagna process," Environmental Science and Technology, vol. 29, no. 10, pp. 2528–2534, 1995.

66. S.-S. Kim, S.-J. Han, and Y.-S. Cho, "Electrokinetic remediation strategy considering ground strate: a review," Geosciences Journal, vol. 6, pp. 57–75, 2002.

67. Start-Ease, Design-Expert: Version 8 Software for Window, Start-Ease, Minneapolis, Minn, USA, 2011.

68. T. Li, S. Yuan, J. Wan et al., "Pilot-scale electrokinetic movement of HCB and Zn in real contaminated sediments enhanced with hydroxypropyl-β-cyclodextrin," Chemosphere, vol. 76, no. 9, pp. 1226–1232, 2009.

69. USEPA, Method 3545: Pressurized Fluid Extraction (PFE), United States Environmental Protection Agency, 1996.

70. USEPA, Method 7000B—Flame Atomic Absorption Spectrophotometry, United States Environmental Protection Agency, 2007.

71. USEPA, Method 7473—Mercury in Solids and Solutions by Thermal Decomposition, Amalgamation, and Atomic Absorption Spectrophotometry, United States Environmental Protection Agency, 2007.

72. USEPA, Method 8270D—Semivolatile Organic Compounds By Gas Chromatography/Mass Spectrometry (GC/MS), United States Environmental Protection Agency, 2007.

73. V. P. Evangelou, Environmental Soil and Water Chemistry: Principles and Applications, Wiley-Interscience, New York, NY, USA, 1998.

74. Y. B. Acar and A. N. Alshawabkeh, "Principles of electrokinetic remediation," Environmental Science & Technology, vol. 27, no. 13, pp. 2638–2647, 1993.

CITATION

Salihu Lukman, Alaadin Bukhari, Muhammad H. Al-Malack, Nuhu D. Mu'azu, and Mohammed H. Essa, "Geochemical Modeling of Trivalent Chromium Migration in Saline-Sodic Soil during Lasagna Process: Impact on Soil Physicochemical Properties," The Scientific World Journal, vol. 2014, Article ID 272794, 20 pages, 2014. doi:10.1155/2014/272794.

CHAPTER 8

SSI on The Dynamic Behaviour of A Historical Masonry Building: Experimental *Versus* Numerical Results

Francesca Ceroni, Stefania Sica, Angelo Garofano and Marisa Pecce

Department of Engineering, University of Sannio, P.za Roma 21-82100 Benevento, Italy

ABSTRACT

A reliable procedure to identify the dynamic behaviour of existing masonry buildings is described in the paper, referring to a representative case study: a historical masonry palace located in Benevento (Italy). Since the building has been equipped with a permanent dynamic monitoring system by the Department of Civil Protection, some of the recorded data, acquired in various operating conditions, have been analysed with basic instruments of the Operational Modal Analysis in order to identify the main eigenfrequencies and vibration modes of the structure. The obtained experimental results have been compared to the numerical outcomes provided by three detailed Finite Element (FE) models of the building. The influence of Soil-Structure Interaction (SSI) has been also introduced in the FE model by a sub-structure approach where concentrated springs were placed at the base of the building to simulate the effect of soil and foundation on the global dynamic behaviour of the structure. The obtained results evidence that subsoil cannot *a priori* be disregarded in identifying the dynamic response of the building.

KEYWORDS

Masonry; Historical buildings; Dynamic monitoring; Eigenfrequencies; Vibration modes; Finite elements models; Soil-structure interaction (SSI)

INTRODUCTION

The structural analysis and the evaluation of seismic vulnerability of existing masonry buildings are often complicated by uncertainty on geometry, typologies and mechanical properties of the materials, effect of age and past loading history, presence of interventions with different building techniques and materials, *etc.* These factors make each masonry structure a unique case, needing a detailed and specific analysis, especially in the case of heritage buildings, whose uncertainties make a reliable prediction of their dynamic behaviour difficult to be carried out.

Recently, ambient vibrations due to both natural and/or artificial sources have been used to directly assess the dynamic behaviour of existing structures in terms of both eigenfrequencies and vibration modes by means of experimental *in situ* measures [1,2,3,4,5,6,7,8,9,10].

The traditional experimental procedures for dynamic identification [2,3,4,11,12], in which a measurable input such as hammer or a shaker is applied to the system and the induced response is later interpreted (input-output identification), are neither feasible nor practical for heritage structures [10]. For this reason, output-only identification methods based on freely available ambient vibrations (from wind, traffic, ground motion, *etc.*) are becoming preferred.

Ambient vibration techniques are not invasive and their use is accepted for heritage buildings since the preservation requirements strongly recommended for this kind of buildings are acknowledged [13]. Moreover, dynamic *in situ* inquiries based on ambient vibrations are not expensive and are, thus, very attractive from an economic point of view for countries like Italy where heritage buildings abound and need to be preserved.

Dynamic *in situ* tests are often used for assessing uncertain structural parameters of masonry (*i.e.*, elastic properties, unit weight) by comparing numerical predictions, coming from detailed Finite Element

(FE) models of the structure, with some experimental evidences [2,11,12,14,15,16,17].

Soil-Structure Interaction (SSI) can be also investigated by interpreting *in situ* dynamic measurements when free-field surface records are available at a nearby site not influenced by building's vibrations [18,19,20]. Unfortunately, there are very few studies on direct identification of SSI from recorded motions [18,21,22]. In most cases, the reference site is, indeed, located too close to the structure or at the ground floor of the building, thus no real free-field conditions occur to experimentally characterize the dynamic response of the foundation soil.

In situ dynamic monitoring of buildings may be performed by temporary or permanent sensor configurations. In the first case, monitoring is not continuous and several test configurations, corresponding to different positions of the recording sensors, may be adopted to achieve a sufficient level of information on the dynamic properties of the building [12]. In the second case, recording is continuous and the sensors are located in fixed positions of the structure. This type of *in situ* dynamic monitoring is primarily conceived as an alert system in case of earthquakes or other types of accidental dynamic sources, but it can be also used for assessing the dynamic behaviour of the structure under environmental actions and for evaluating eventual damage level of a structure after an earthquake [23].

In this paper, the detailed study of a historic masonry building (Palazzo Bosco Lucarelli) located in Benevento (Italy) is presented focusing the attention on the identification of the main eigenfrequencies and vibration modes by means of the experimental data provided by the *in situ* dynamic monitoring of the structure. The experimental outcomes are compared to the results of numerical modal analyses performed by three different 3-dimensional FE models of the building aimed to investigate the influence of SSI on its dynamic behaviour. The building has been firstly modelled as completely restrained at the basement while, in a second stage, SSI has been investigated by placing rotational and translational springs at the base of the ground floor. Compared to former studies carried out by the authors on the same case-history [24,25], the present paper highlights the following issues: (1)

comprehensive interpretation of the available experimental measures to detect both frequencies and vibration modes of the building by means of cross-spectra, coherence and phase-angle functions between different signals; (2) evaluation of SSI effects on the dynamic behaviour of the building.

THE CASE STUDY

The case study herein examined has been widely described in [24]. It is a masonry building, called "Palazzo Bosco Lucarelli", located in Benevento (Italy); the current structure (Figure 1a) has a rectangular holed plan and consists of an underground floor, a ground floor, two upper levels, and an attic under the pitched roof. The largest dimensions in plan are about 33 m and 26 m and the total height is 18.2 m. The thickness of the walls varies in the range from 0.60 m to 1.30 m.

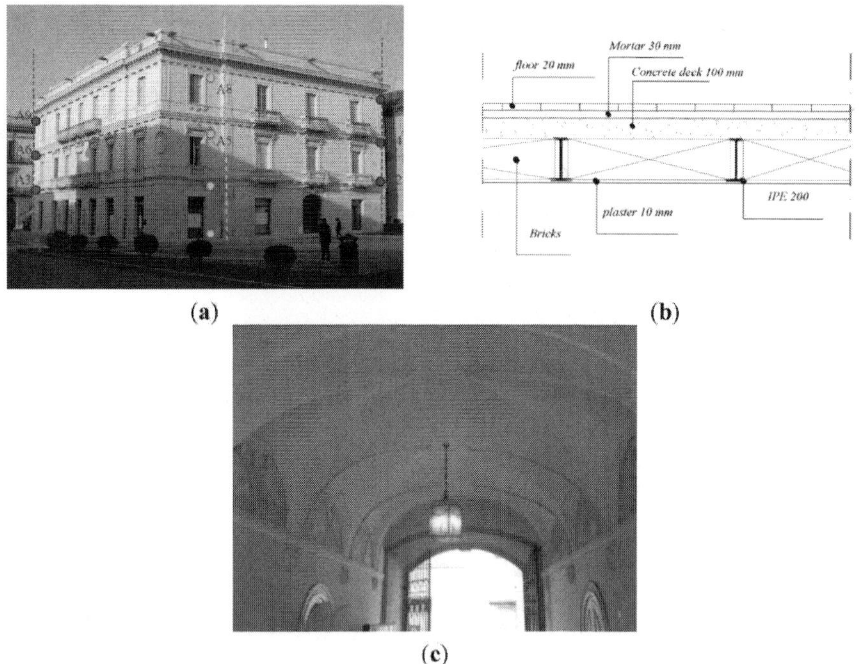

Figure 1. Palazzo Bosco Lucarelli: (**a**) Current photo of the building with instrument locations; (**b**) Schematic draw of the typical floor; (**c**) Vaults of the closed court at the ground floor.

The *in situ* survey detected a masonry structure made of different materials and textures at the three levels, due to several construction phases. At the underground and ground level, walls are made of irregular blocks of limestone and conglomerates. At the higher levels, the walls are made of clay bricks covered by a reinforced cement plaster (thickness 50 mm with a grid of steel bars diameter 6 mm spaced of 150 mm). More details about the geometrical configuration, the material typologies and the assessment of the mechanical properties of masonry are reported in [24].

The floors are made of standard steel profiles with height of 200 mm (IPE 200), spaced of 85 mm and interspersed with hallow brick tiles and covered by a concrete deck 100 mm thick (Figure 1b). The ceilings of the staircase and of some rooms in the entrance hall at the ground floor (Figure 1c,b) are made of masonry vaults. The roof is made of a steel truss covered by a profiled steel sheeting and brick tiles.

Based on the Down-Hole test (DH), carried out very close to the building, a softer layer with an average shear wave velocity $V_s = 600$ m/s has been detected in the first 12 m of the subsoil interacting with Palazzo Bosco. More details on the geotechnical characterization of the site have been reported in [24]. Linear site response analyses, specifically carried out, show that the fundamental frequency of the subsoil below Palazzo Bosco is quite high (around 10 Hz), as typical of relatively stiff soils.

EXPERIMENTAL STUDY OF THE DYNAMIC BEHAVIOUR OF THE BUILDING

Methodology for Data Treatment

Structural dynamic identification methods can be grouped in two main categories: analytical and experimental.

In the analytical approaches, starting from the knowledge of structural geometry, initial conditions, characteristics of the materials, distribution of mass, stiffness, and damping of the structure, an eigenvalue problem is solved to determine the dynamic parameters of the system (frequencies and modal shapes). The assessment of the dynamic behaviour of the structure is clearly dependent on the knowledge level available for the above mentioned parameters.

The experimental approaches may be divided in "traditional" or "operational" modal analysis. In the former, starting from the measurements of the dynamic input and of the structural response, the transfer function is calculated using the experimental modal analysis (EMA). The method is also known as "inverse problem" [10], or "input-output identification" in which the input (excitation) and the output (structural response) are both known and allow the dynamic parameters of the structure to be assessed independently of the knowledge level of geometry, mass distribution and material properties of the structure.
The operational approach, conversely, is an "output-only identification" method, which, starting from the structural response (output) without knowing the input force, allows estimating the dynamic parameters of the structure using the instruments of Operational Modal Analysis (OMA) [1,4,10], also in this case independently of the detailed knowledge of the structure.

The simplest OMA technique is the Peak Picking Method (PPM) where Fourier transforms of the recordings are carried out and frequency peaks are detected on the Fourier amplitude spectra. If the mode shapes of the structure are well separated, the PPM may provide a rough indication of the dynamic parameters of the structure. When the modal shapes of the structure are closer to each other, much more refined analysis methods should be applied. Worth mentioning is the Frequency Domain Decomposition (FDD) technique, based on the decomposition of the Power Spectral Density (PSD) matrices into single-degree-of-freedom systems [1,3,4,26,27].

In this paper, an output-only identification is adopted using the basilar instruments of the OMA for identifying the main frequencies and vibration modes of the examined building. After a preliminary baseline correction, the accelerograms acquired on Bosco-Lucarelli Palace were interpreted in the frequency domain by basic spectral analysis by computing cross-spectrum, correlation and phase functions between different couples of recorded signals. In the following section, the available experimental measures will be discussed in more detail with respect to former works [24,25].

Instrumentation Set-Up

As detailed in [25], since 2008, the building has been equipped by the Italian Department of Civil Protection (DPC) with a permanent monitoring system, working as a warning system to medium-high seismic events. Even if the instrumentation system was not conceived for the dynamic identification of the structure, the measures obtained during the trial phase of the system were sufficient to identify the main frequencies and vibration modes of the building.

Nine couples of mono-axial accelerometers were placed at the three floors of the building and a three-axial accelerometer was located at the ground floor (Figure 1a and Figure 2a). Each couple of sensors is able to be measured separately along the X and Y direction.

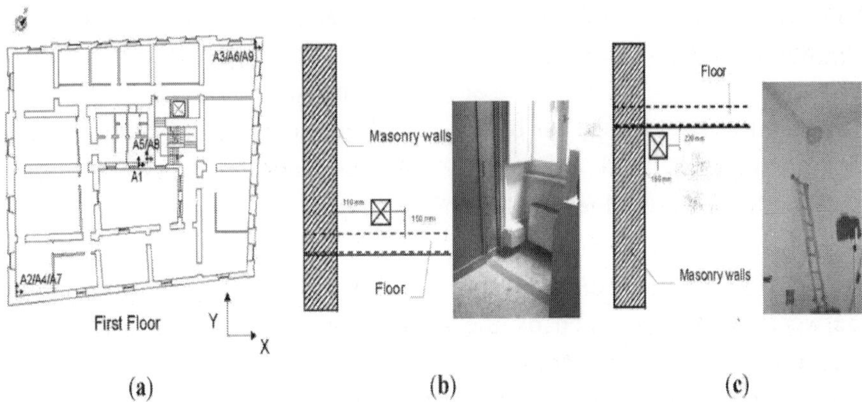

Figure 2. (a) Plan of the first floor with instrument locations; (b) Example of position of the accelerometers at the 1st floor (A1, A2, A3), and (c) at the second (A4, A5, A6) and third floor (A7, A8, A9).

The sensors have been aligned as accurately as possible along three vertical lines: one corresponds approximately to the centre of the building and the other two to the external edges. The central vertical line is made of sensors A1, A5, A8, while the two external corners of sensors A3, A6, A9, and of A2, A4, A7 (Figure 1a and Figure 2a). The couples A1, A2, A3 are placed at about 100 mm from the first floor intrados (Figure 2b), the couples A4, A5, A6 at about 100 mm from the second floor extrados (Figure 2c), and the couples A7, A8, A9 at about 100 mm from the third floor extrados (Figure 2c).

Different sets of raw accelerometric recordings were provided by DPC to the authors who carried out their own elaboration of the experimental measures and developed FE models independently of the activities of DPC [28]. These recordings were caused by ambient noise and impulsive sources, both measured during the installation of the monitoring system and the trial stage and assumed as unknown dynamic input.

Dynamic Measures under Impulsive Sources and Environmental Actions

During the installation and the trial stage of the monitoring system, an impulsive source was activated by the fall of a concrete block (about 1 m^3) on a truck placed close to the building and several measures under environmental noise have been acquired. The same methodology of data treatment has been used for both types of measures by computing cross-spectrum, phase and coherence functions between pairs of signals recorded at different locations of the building.

Figure 3 shows, indeed, cross-spectrum, coherence and phase functions between signals recorded by the couples of sensor A3–A9 during the concrete block fall in direction X and Y, respectively. Cross-spectra were computed between signals recorded at different elevations (first floor for A3 and third floor for A9) along a single vertical.

Analogously, Figure 4 shows the same graphs for the couple of sensor A1 and A8 (first floor for A1 and third floor for A8) along the central vertical of the building.

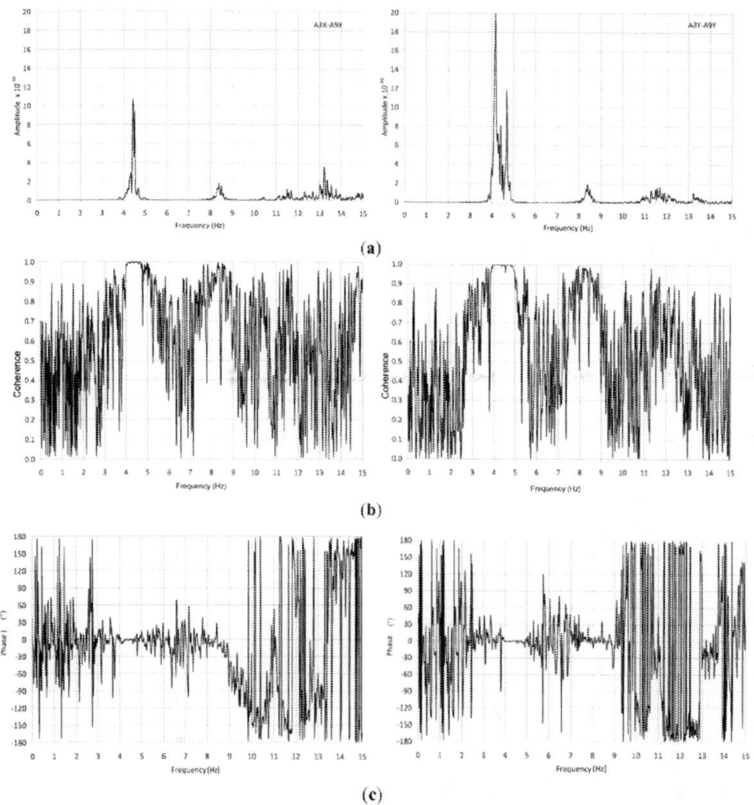

Figure 3. Signals registered by sensors A3 and A9, induced by the concrete block fall in X and Y direction: (**a**) cross-spectra; (**b**) coherence; and (**c**) phase angle.

The results highlight clearly a first frequency in Y direction at about 4.2 Hz and a first frequency in Xdirection at about 4.7 Hz (Figure 3a and Figure 4a), since the coherence function is close to one in correspondence of such frequencies (Figure 3b and Figure 4b). Since at the above two frequencies the phase difference (Figure 3c and Figure 4c) between the selected pairs of sensors is zero and the same happens also for pairs of points on the opposite side of the structure but at the same elevation (e.g., between sensors A7 and A9), this means that these two modes are not coupled and are primarily translational. Moreover, it is worth noting that in both directions the experimental measures show at least two peaks in the range 4–5 Hz, even if only one is predominant.

A second frequency in X direction may be clearly identified by all sensors at around 8.4 Hz (coherence function close to one as can be seen in Figure

3b and Figure 4b). The level of amplitude registered for such peaks, however, is lower than the ones registered for the first frequency.

In the Y direction, only a few sensors evidence a second smaller peak in the range of frequencies 8.0–8.5 Hz. This latter cannot be considered as an effective frequency of the building. However, the reduced amplitude of the cross-spectrum indicates that the participating mass associated to these higher modes, if they effectively exist, is negligible.

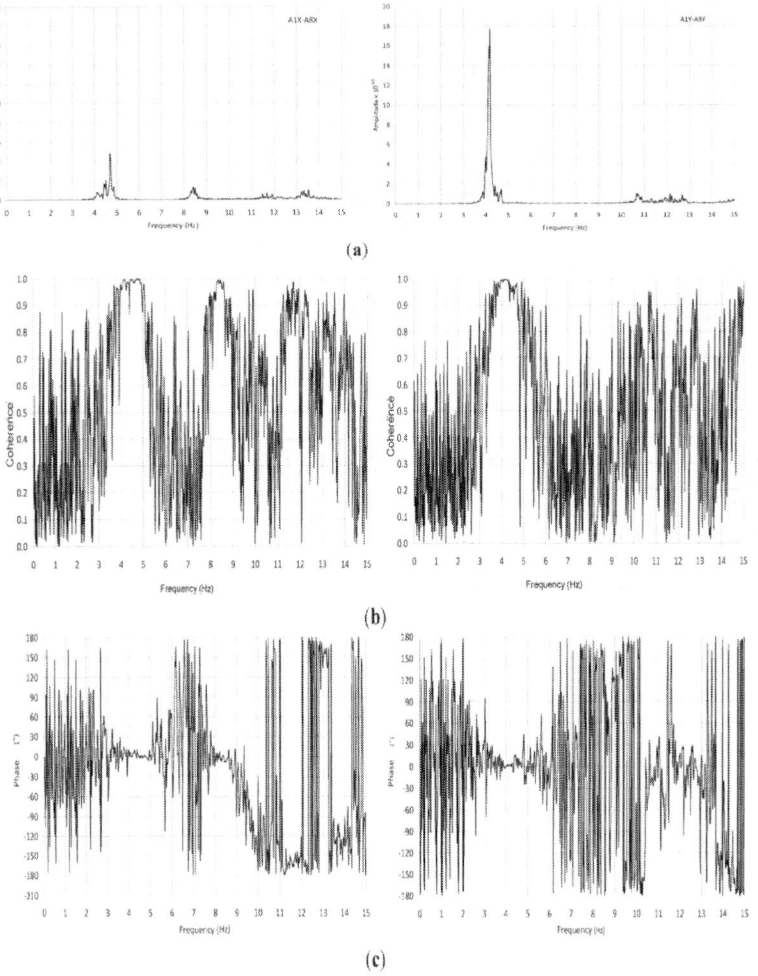

Figure 4. Signals registered by sensors A1 and A8, induced by the concrete block fall in X and Y direction: (**a**) Cross-spectra, (**b**) Coherence; and (**c**) Phase angle.

Finally, most sensors evidence higher-order frequencies in the range 11–14 Hz, both along X and Y direction. An increased level of noise is, however, detectable for such frequencies since much lower values of the coherence function have been found and, thus, the identification process becomes more uncertain.

With the same methodology described above, also the recordings due to environmental noise have been interpreted. For the sake of brevity, in Figure 5 only the cross-spectra for the three monitored verticals are reported. The cross-spectra here shown were computed between signals recorded at the first and third floor of each vertical, in direction X and Y. The first two frequencies in Y (4.2 Hz) and X direction (4.7) are again found and identified very reliably, based on values of coherence and phase that were found to be equal to 1 and 0, respectively. Further peaks can be observed in the ranges 8.0–8.5 Hz and 11.0–11.5 for the X direction and, in a less pronounced way, also along Y. Again, these frequencies, even if they correspond to real modes of the structure, can be considered not very significant due to the low amplitude of the cross-spectra.

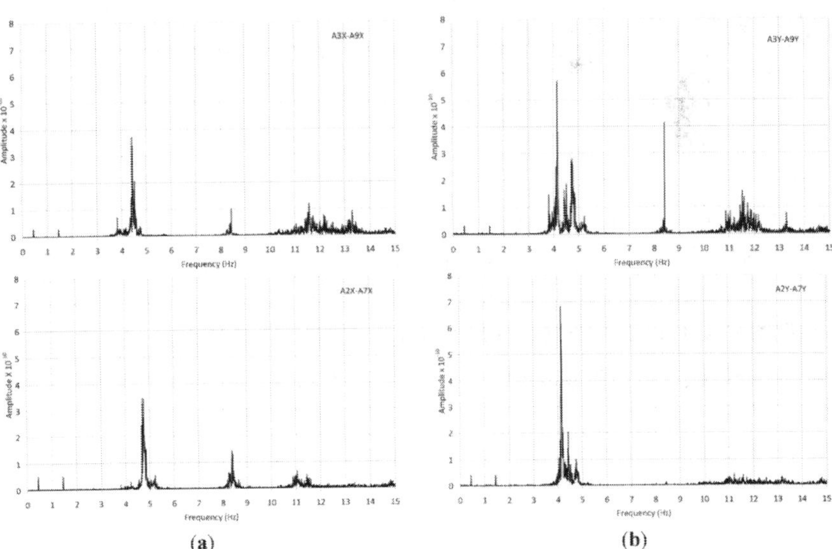

Figure 5. From top to bottom: cross-spectra between signals registered by sensors A3–A9 and A2–A7, induced by environmental noise, along (**a**) X and (**b**) Y direction.

In Table 1, a synthesis of the two main frequencies identified by the different type of recordings is listed. The comparison highlights a substantial uniformity in the values of the eigenfrequencies which were experimentally identified under different dynamic sources, as evidenced by the very low values of Coefficient of Variation (CoV). All experimental analyses evidence, indeed, that the frequency in the Y direction (first mode) is slightly lower (about -10%) than the one related to the X direction (second mode). This can be indicative of a greater deformability of the building in the direction Y. To corroborate this assumption it can be observed that along the shorter side (direction Y) the building has a lower amount of masonry walls, often not continuous, than in the orthogonal direction X (see plans of the building in Figure 2a).

Table 1. Main eigenfrequencies of the building experimentally identified by the sensors under different dynamic sources. CoV: Coefficient of Variation.

Mode	Direction	Frequencies (Hz)					
		Environmental registration 1	Environmental registration 2	Fall of concrete block	Mean (Hz)	Stardard deviation (Hz)	CoV (−)
I	Y	4.2	4.4	4.2	4.3	0.14	3%
II	X	4.6	4.8	4.7	4.7	0.10	2%

The experimental vibration modes of the building corresponding to the identified frequencies have been also obtained. They will be discussed in the following section through comparison with the numerical predictions obtained from different FE models of the building.

FINITE ELEMENT MODELLING OF THE BUILDING: LINEAR MODAL ANALYSIS

Soil-Structure Interaction by Means of Elastic Concentrated Springs

The dynamic behaviour of a structure can be modified by the presence of a deformable soil. As well-known, a flexibly-supported structure has a fundamental period longer than the period T of the corresponding fixed-base structure, and a higher damping due to energy dissipated into the soil through wave radiation. This latter phenomenon is inhibited in rigidly-supported structures.

The influence of the subsoil on the main frequencies of the building depends on features of the soil, generally synthesized by the value of shear wave velocity, V_s and of the structure, generally represented by its stiffness and height. In literatures [29,30,31], several analyses evidenced that the influence of SSI on the dynamic behaviour of a structure essentially depends on the relative stiffness (and mass) of the superstructure compared to that of the soil interacting with the building. In short, the effects of SSI can be relevant for massive and stiff superstructures compared to the foundation soil and also for slender and tall buildings, like towers, in very soft soils [32].

In the case of Palazzo Bosco, the subsoil interacting with the building is enough stiff but, in principle, not a rock (conventionally, in geotechnical earthquake engineering it is considered "rock" a material with $V_s > 800$ m/s), as the first 12 m below the ground level have lower values of V_s (in Section 2 an average value of 600 m/s has been estimated).

A traditional approach for the assessment of SSI in the dynamic field consists in introducing to the structural model of the building the effect of both soil and foundation system by means of concentrated rotational and translational springs placed at the structure base, which becomes in this way no more fixed. These springs are complex functions (impedances) whose real part (stiffness) depends on soil shear stiffness, G, and Poisson ratio, υ, foundation geometry, and frequency of excitation, ω [30].

In general, the impedance functions k_i depend on frequency ω of the input motion and can be expressed multiplying the static stiffness, K_i, by a frequency-dependent coefficient, α_i [30]:

$$k_i(\omega)=\alpha_i(\omega)\cdot K_i \qquad (1)$$

For the estimation of the foundation impedances, in [30] different analytical expressions and/or charts are provided, depending on the shear stiffness, G, and the Poisson coefficient, v, of the soil, the geometrical characteristics of the foundation (and the frequency, ω of the input motion. In [31], a state-of-the-art about different formulas proposed in literature for the evaluation of SSI is reported, taking into account: (1) the shape of the foundation (circular, strip, rectangular/square foundation); (2) the soil model (homogeneous half-space, soil layer over rock); and (3) the foundation embedment (foundation placed on the limit surface of the half-space or embedded in it).

It is worth noting that the available closed-form solutions have been all obtained assuming a linear elastic response of the foundation soil. This assumption on soil behaviour is reliable when low levels of strain are involved, as in the case of vibrations due to low-amplitude excitations (environmental actions). For strong earthquakes, however, soil non-linearity should be accounted for, which induces a decrease in soil stiffness with respect to the initial value G_0 (derived from the shear wave velocity V_s).

The Finite Element Models
The examined building has been numerically studied by a three-dimensional Finite Element model made of shell elements (software SAP2000, release 14, CSI, Berkeley, CA, USA). The three floors have been modelled by rigid connection between all the nodes belonging to each plane, which are constrained against in-plane deformations (hypothesis of rigid floor).

The vertical loads were defined with respect to the actual loading conditions of the building at the moment of the trial tests. For the permanent and variable load, respectively, the values G_k = 5.00 kN/m^2 and Q_k = 0.25 kN/m^2 have been assumed. The low value of the considered variable load is due to suspension of teaching activities in the classrooms during the dynamic testing. The dead load of the steel roof has been estimated to be 1.75 kN/m^2. All the vertical loads related to the floors have been directly applied on the top of the masonry walls supporting the floors.

Based on the values of unit weight of the single blocks and on the masonry texture and considering the presence of the reinforced plaster layer, for the first and second floors the unit weight of the masonry has been estimated to be about 20 kN/m^3. Analogously, for the irregular masonry of the ground and underground floor, the unit weight has been assumed as 24 kN/m^3. More details about the assessment of unit weight can be found in [24].

As no experimental data are available to characterize the masonry Young's moduli, the indications of the applicative Italian code [33,34] (Appendix 8, Table C8A.2.1) were firstly adopted in previous analyses [24] and then modified. The estimation of the Young's modulus is, indeed, an open problem since the uncertainty and variability of E, also within the same type of stone, can be usually very high (CoV can be also greater than 40% [35]). Detailed sensitivity analyses about the effect of E on the numerical frequencies of the examined building in the linear field have been already carried out in [24] and evidenced that a variation of E of 50% leads to a variation of frequencies of about 25%. Since in this paper, attention is mainly focused on the assessment of soil deformability on the dynamic behaviour of the examined building, in the analyses here reported the Young's modulus of masonry has not been varied. In particular, the values already established in [24] have been assumed: E_G = 1850 MPa and $E_{1\text{-}2}$= 1550 MPa for the ground level and for the upper levels, respectively.

Three FE models have been considered as the base restraints change in order to compare the experimental *vs.* numerically predicted behaviour of the buildings both in terms of main frequencies and vibration modes and check the effect of SSI.

In the first model, the underground level of the building is neglected and the ground floor is considered completely restrained at the base (fixed-base model).

In the second model, the underground level of the building has been added in the model assuming for the masonry walls a material having the same characteristics of the walls of the ground floor; its base (placed 3 m under the ground level) has been fully constrained as in the first model. Moreover, in this model the effect of soil around the walls of the underground level was neglected.

The FE model without the underground level is made of 16,895 elements, while the model with the underground level is made of 20,477 elements. In both cases, the maximum area of the finite elements is 0.25 m^2.

Finally, a third model has been considered, in which the underground level has been neglected and a set of six concentrated linear springs (three translational and three rotational) has been imposed at the base of the ground floor of the building for taking into account Soil Structure Interaction. In this case, the nodes of the FE model belonging to the base of the building have been constrained in order to reproduce the effect of a stiff plate with respect to the in-plane and out-of-plane differential deformations. The concentrated springs have been positioned at this level, in correspondence to the centre of mass of the building. The spring values (*i.e.*, the foundation impedances) have been computed according to the analytical expressions of [30], considering the scheme of a rectangular foundation embedded in an elastic half-space. The following parameters have been adopted for the foundation soil: density ρ = 2000 kg/m^3, Poisson's ratio v = 0.3, shear waves velocity V_s = 600 m/s. This latter parameter—as stated above–represents the average value of V_s in the significant soil volume interacting with the building. In the third model, the underground level of the building has not been modelled, but its presence has been however taken into account in the spring stiffness values, since the foundation depth, D, below the ground level and the height, d, of the sidewall in effective contact with the surrounding soil have been assumed both equal to 3 m, that is the height of the underground level.

The use of such a modelling approach has been found to be more suitable than modelling also the underground floor and applying at its base the springs. In fact, the effect of confinement due to the lateral soil is introduced in the model by taking into account the foundation embedment.

Numerical Dynamic Behaviour of the Building
The dynamic behaviour of the building in the elastic field has been numerically investigated by the 3D FE models described above by means of linear modal analyses.

In Table 2 periods, frequencies, and modal participating mass ratios associated to the first seven (in order to achieve a total mass participant at least 75% in both directions X and Y) vibration modes of the building are reported.

Table 2. Period, frequencies and modal participating mass ratios for the fixed-base Finite Element (FE) model of the building.

| Mode | Period | Frequency | Translation (−) | | | Rotation (−) | | |
	T (s)	f (Hz)	X	Y	Z	RX	RY	RZ
1	0.213	4.70	0.00	**0.70**	0.00	0.30	0.00	0.26
2	0.195	5.13	**0.38**	0.01	0.00	0.00	0.11	0.45
3	0.173	5.79	**0.31**	0.00	0.00	0.00	0.09	0.01
4	0.081	12.36	0.00	**0.04**	0.00	0.01	0.00	0.02
5	0.077	12.97	0.00	**0.07**	0.00	0.04	0.00	0.02
6	0.076	13.15	**0.05**	0.00	0.16	0.11	0.00	0.03
7	0.072	13.89	0.02	0.00	0.51	0.24	0.42	0.02
Sum participating mass			0.76	0.82	0.67	-	-	-

The first eigenfrequency of the building is related to a vibration mode with a participating mass ratio of 70% and is completely translational in Y direction as it can be evidenced by the deformed shape depicted in Figure 6a. The deformed shapes of the building corresponding to the second and third modes (seeFigure 6b,c) show a torsional shape of the building, with a prevalent translational component in Xdirection as evidenced by the participating mass ratios of 38% and 31%, respectively.

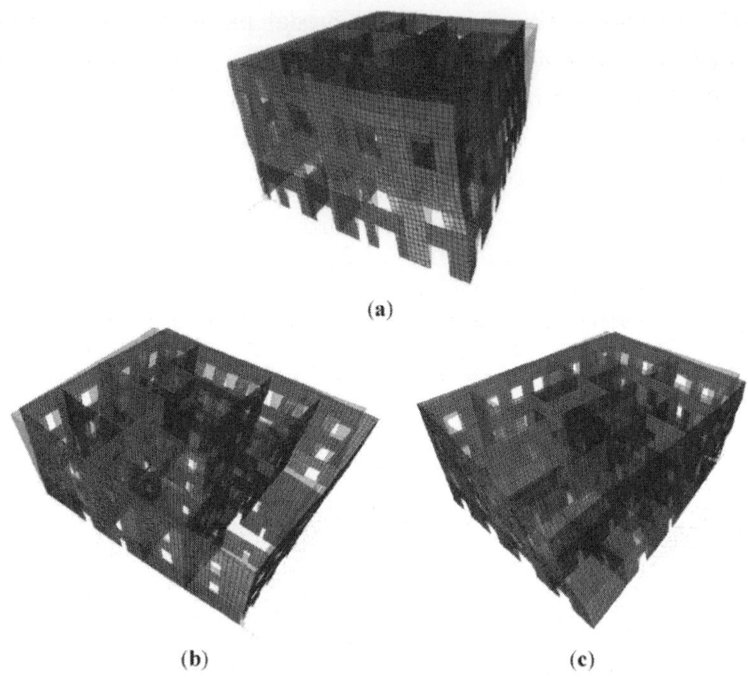

Figure 6. Main vibration modes of the structure in the fixed-base FE model: **(a)** First mode; **(b)** Second mode; **(c)** Third mode.

As highlighted by the experimental results too, the numerical frequency in Y direction (4.7 Hz) is slightly lower (about −10%) than the second and third frequencies characteristic of the X direction (5.13 and 5.79 Hz), that confirms the tendency of the building to be more deformable in the direction Y parallel to the shorter side (see plan in Figure 2a).

The higher numerical modes are characterized by very low participating mass ratios and the deformed shapes of the building show that they are local modes, since they regard single walls.

When the underground level is added to the fixed-base model, the deformability of the structure clearly increases, with an increment of about 16% and 12% in the first vibration period along the Y and X direction, respectively, as reported in Table 3 for the second model.

The first three vibration modes are similar to the ones of the fixed-base model (model 1), even if some differences in the mass participating can be observed (*i.e.*, the mass participating in X direction increases in the second mode and decreases in the third). Moreover, a fourth and a fifth mode in Y and X direction, respectively, are better identified in this model, but they are characterised by a low amount of mass participating (15% and 12%) and, thus, they can be considered as local modes.

Table 3. Period, frequencies and modal participating mass ratios for the FE model of the building with the underground floor.

Mode	Period T (s)	Frequency f (Hz)	Translation (−) X	Y	Z	Rotation (−) RX	RY	RZ
1	0.249	4.02	0.00	**0.66**	0.00	0.25	0.00	0.26
2	0.220	4.55	**0.44**	0.00	0.00	0.00	0.11	0.37
3	0.196	5.10	**0.21**	0.00	0.00	0.00	0.06	0.02
4	0.094	10.65	0.00	**0.15**	0.00	0.04	0.00	0.06
5	0.088	11.40	**0.12**	0.00	0.03	0.03	0.01	0.06
6	0.083	12.02	0.00	0.00	0.07	0.07	0.03	0.00
7	0.083	12.08	0.01	0.00	0.54	0.26	0.34	0.01
Sum participating mass			0.78	0.81	0.64	-	-	-

For the third model, characterized by the concentrated springs placed under the ground level, periods, frequencies, and modal participating mass ratios associated to the first seven vibration modes, are reported in Table 4. The overall effect of the soil–foundation interaction can be assumed as comprehensive of the part of the building that interacts with the soil both laterally and under the structure.

Table 4. Period, frequencies and modal participating mass ratios for the FE model with concentrated springs

Mode	Period	Frequency	Translation (−)			Rotation (−)		
	T (s)	f (Hz)	X	Y	Z	RX	RY	RZ
1	0.224	4.45	0.00	**0.72**	0.00	0.32	0.00	0.27
2	0.203	4.92	**0.50**	0.00	0.00	0.00	0.15	0.41
3	0.179	5.57	**0.21**	0.00	0.00	0.00	0.07	0.03
4	0.090	11.07	0.00	0.00	**0.86**	0.46	0.44	0.00
5	0.085	11.72	0.00	**0.08**	0.00	0.01	0.00	0.04
6	0.082	12.18	0.00	**0.04**	0.00	0.03	0.00	0.01
7	0.081	12.39	**0.08**	0.00	0.02	0.00	0.17	0.04
Sum participating mass			0.79	0.84	0.88	-	-	-

The first vibration mode shows again a well-defined translational modal shape in Y direction (Figure 7a) with a participating mass ratio (72%) lightly higher than the value achieved in the fixed-base scheme (70%, fixed-base Model 1). As in the previous models, the second and third modes evidence a slightly torsional shape of the building, with a prevalent translational component in X direction (Figure 7b,c), even if for the second mode the participating mass ratio in X direction further increases (50% *vs.* 38% of the fixed-base Model 1), while in the third mode it reduces (21% *vs.* 30%).

The presence of springs leads to a reduction of the frequencies, *i.e.*, an increase of the period of about 5% and 4% for the first and the second vibration mode, respectively, compared to the values predicted by fixed-base Model 1.

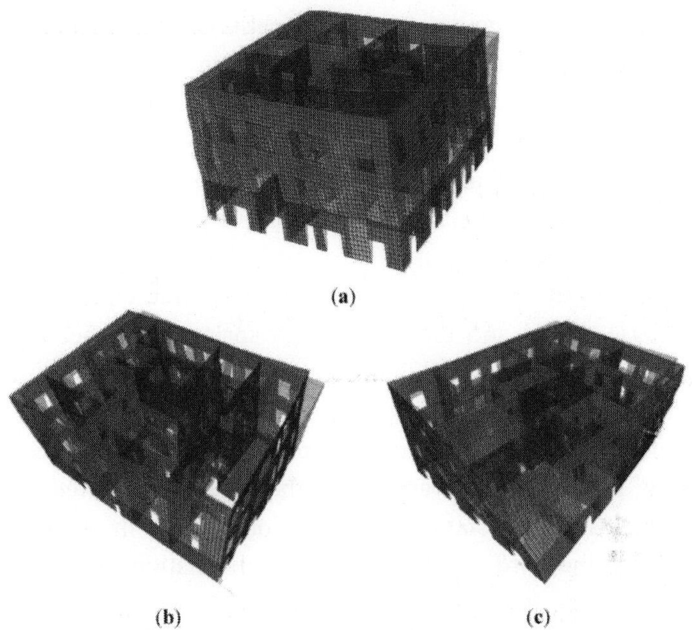

Figure 7. Main vibration modes of the structure in the FE model with concentrated springs at the base: **(a)** First mode; **(b)** Second mode; **(c)** Third mode.

A singularity of the dynamic behaviour of the building modelled with springs is that the fourth mode is characterized by a vertical deformation along the Z direction with a high ratio of participating mass (86%); this mode is clearly caused by the presence of the springs that introduce the possibility of a translation in the Z direction.

The fifth and the sixth mode show again a very low participating mass along the Y direction and, thus, they can be considered as local.

In Figure 8, the first vibration modes in Y direction given by the fixed-base model and the model with springs are compared considering the displacement along the Y direction of the vertical lines at the corners. Since the building has a rectangular plan and is not very tall, the modal shapes cannot be represented by only one vertical and, thus, the deformations along the four corners of the building were plotted. In particular, in Figure 8, the two lines for each model (black lines for the fixed-base and grey lines for the model with SSI) represent the envelope lines (minimum and maximum displacements) of the numerical modal shapes evaluated at each corner of the building. In the model with the springs, the displacements are not zero at the basement due to the presence of the springs; this initial offset dampens along the height and, however, the modal shapes furnished by the two models are quite similar.

To have more exhaustive comparisons, the first three vibration modes of the fixed-base model and the model with springs have been compared with respect of the positions identified in the plan of the building by the four corners of the structure. In particular, the original and the deformed position of the top of the structure is represented in the pictures reported in Figure 9. Figure 9 shows again that the modes associated to the first two frequencies are translational in direction Y for both models and that the second and the third modes are not only translational in direction X.

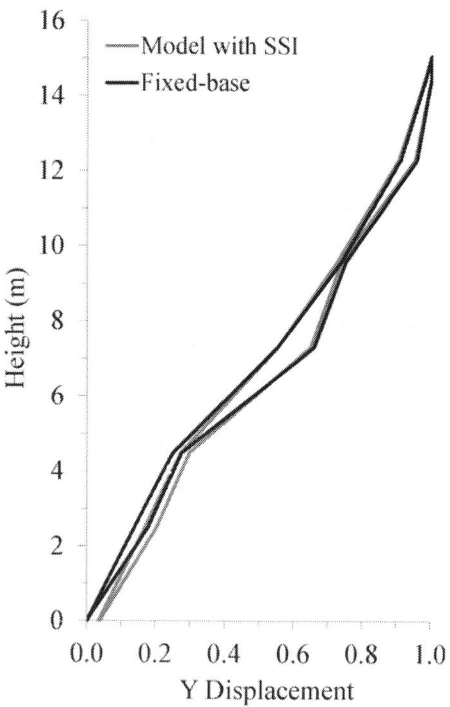

Figure 8. Comparison between the first modal shapes (Y direction) at the corners of the building given by the fixed-base model and the model with concentrated springs at the base.

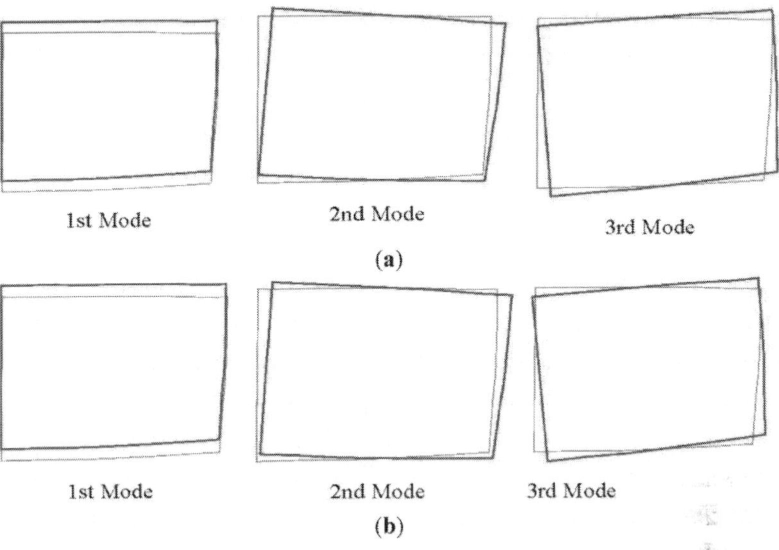

Figure 9. Comparison between the first three numerical vibration modes of the structure given by: **(a)** The fixed-base model; **(b)** The model with concentrated springs at the base.

Comparison of Linear Dynamic Analysis with Experimental Results
The comparison of the first experimental frequency (4.3 Hz in Y direction) with the first numerical one given by the fixed-base model (4.70 Hz) shows a difference of about 10%. This difference reduces to only 3% in the model with the springs (4.45 Hz) pointing out that there could be an effect of the SSI in making the building more deformable and, thus, more similar to the experimental behavior. Clearly, this is true, if the elastic properties and unit weight of the materials and geometry and loads acting on the structure have been correctly assessed.

About the X direction, all the FE models have provided two frequencies (the second and the third) very close to each other (*i.e.*, 5.13 and 5.79 Hz for the fixed base model and 4.92 and 5.57 Hz in the FE model with springs) and characterized by different participating mass (38% and 31% for the Model 1 and 50% and 21% for Model 3 with springs). The second experimental frequency in X direction (4.7 Hz) is comparable with the values of the second numerical frequency in X direction and, as occurred for the Y direction, is again closer to the value predicted by the model with springs (4.92 Hz). On the contrary, the third numerical frequency in X direction does not correspond to any experimental values and could be related to type of finite element used in the modelling strategy. Similar linear dynamic analysis carried out by a different FE model made of brick elements instead of shell [24] furnished, indeed, slightly different results, since only one frequency was individuated in X direction corresponding to a pure translational vibration mode very similar to the one detected in Y direction and with the same participating mass. This result, joined to the regularity of the building and to the results of the *in situ* dynamic tests that allowed identifying only two clear uncoupled translational modes in direction Y and X (phase equal to 0), would confirm that the third numerical frequency is not realistic, but has to be attributed to the second mode. Moreover, a confirmation of this hypothesis is the fact that the sum of the participating mass for second and third modes in X direction is comparable with the mass participating associated to the first frequency in Y direction (71% *vs.* 72% for the model with springs).

In Section 3.3, it was observed that only the first two frequencies can be reliably identified by means of the experimental measures, since both in X and Y direction higher order frequencies individuated in the ranges 8.0–8.5 Hz were characterized by low amplitude of cross spectra and lower performing values of coherence and phase. This uncertainty about the effective existence of vibration modes of the building associated to higher order frequencies is also confirmed by the absence of a clear correspondence of such frequencies in the numerical results. However, if a third experimental mode really exists at a frequency ranging in 8.0–8.5 Hz, it could be significant of some local phenomena occurring in the real behaviour of the building which are not reproduced in the FE model. Such local phenomena, as for example the lack or the ineffectiveness of connections between walls and floors, could be detected, and thus reproduced in the model, if a detailed relief of the damage state of the building has been done, but for the current level of knowledge of the building none of these phenomena has been evidenced.

The experimental frequencies detected in X and Y directions in the ranges 11–14 Hz, especially in the measures acquired under environmental conditions, could be associated in the model with springs to the fifth mode associated to the frequency of 11.7 Hz and a mass participating in Y direction of about 8% and to the seventh mode associated to the frequency of 12.4 with a mass participating in X direction of about 8%. However, for such modes the reduced amplitude of the cross-spectra experimentally observed and the low amount of mass participating given by the models lead to considering both of them as negligible.

In conclusion, the comparison between the frequencies assessed experimentally with the numerical values predicted by the FE models has evidenced that only the first two modes can be reliably identified. Moreover, considering the experimental uncertainty and the influence of the FE model adopted, a good agreement between experimental and numerical results has been achieved for the chosen values of elastic properties of the materials and of mass acting on the structure and taking into account the SSI.

In Figure 10 and Figure 11, the first and the second experimental vibration modes of the structure are depicted, respectively, and compared with the numerical ones given by the fixed-base model and the model with springs. In particular, the normalized displacement of the three instrumented vertical lines have been considered, namely corresponding to the central position (sensors A1–A5–A8), the northern (sensors A3–A6–A9) and the southern (sensors A2–A4–A7) corners of the building. The experimental displacement measured in correspondence of the first and second experimental frequencies (4.3 Hz in Ydirection and 4.7 Hz in X direction) have been normalized to the measure of the sensor placed at the top (A8, A9, and A7 for the three vertical lines, respectively). The experimental displacements of first vibration mode measured by the sensors of each vertical in Y direction have been directly compared inFigure 10 with the theoretical ones given by the two FE models, since the first mode is completely translational in Y direction. On the contrary, in Figure 11, for the second mode, the experimental displacements measured in X direction have been compared with the component in X direction of the displacements given by the two FE models, since the numerical second mode was not completely translational in X direction, as previously discussed.

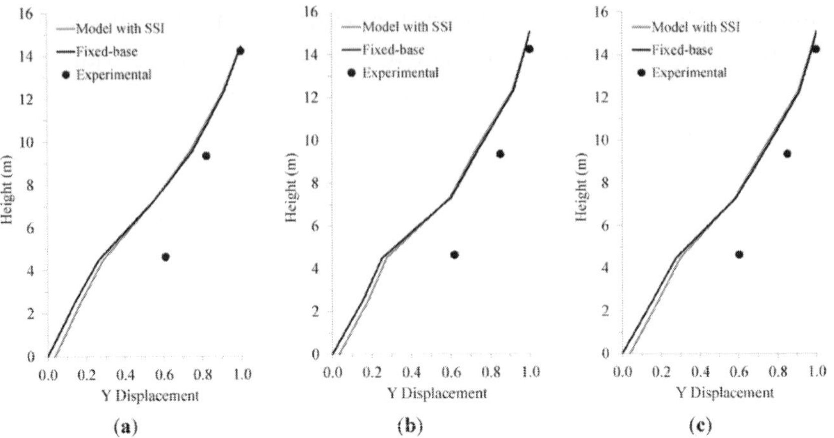

Figure 10. Experimental *vs.* numerical first vibration modes in *Y* direction for: **(a)** The central line (A1–A5–A8); **(b)** The northern corner (A3–A6–A9); and **(c)** The southern corner (A2–A4–A7).

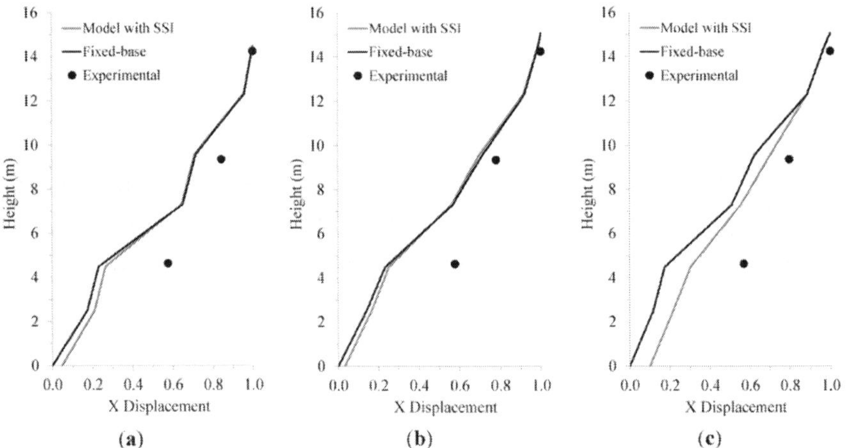

Figure 11. Experimental *vs.* numerical second vibration modes in *X* direction for: **(a)** The central line (A1–A5–A8); **(b)** The northern corner (A3–A6–A9); and **(c)** The southern corner (A2–A4–A7).

The numerical vibration modes are somewhat more consistent than the experimental ones for both frequencies, especially for the displacement measured at the first floor. The larger deformability introduced by the springs placed at the ground floor only leads to a displacement shift at the basement in the model with SSI, while the shape for height is quite similar.

Notwithstanding the small difference in modal shape and frequencies (only 10%) obtained for the case at hand, the model with SSI is surely a better representation of the structure, since it allows considering the effect of the subsoil which contributes to the real dynamic behaviour of any building. The examined case study is, indeed, representative of a complete approach which accounts also for SSI in the modelling.

Other types of structure, such as tower or very deformable buildings, can be even more susceptible to SSI and it is important to model this phenomenon in evaluating their dynamic behaviour [32,36].

Finally, it is worth noting that further dynamic analyses carried out with the FE model with springs evidenced that as the stiffness of the springs reduces, the second vibration mode becomes more regular and shows a well-defined translational shape also in X direction, similarly to the first vibration mode in Ydirection. In addition, the third torsional mode disappears and is substituted by the translational mode along the Z direction which, for the adopted value of spring stiffness, is the fourth one (see Table 4).

In conclusion, the analysis of the vibration modes confirms the necessity of introducing the SSI to have a more reliable model. Such a model may be later used for further dynamic and non-linear numerical analyses.

CONCLUSIONS

The structural analysis of historical masonry buildings is a complex matter, since each construction is a stand-alone system, designed and erected for being a singular case. Dynamic *in situ* tests combined with continuous monitoring of the building is an innovative, effective and non-invasive procedure able to assess the overall dynamic behaviour of the structure in terms of main vibration modes and eigenfrequencies.

For the case study herein presented, starting from the interpretation of experimental data coming from a permanent dynamic monitoring system installed on the building by the Italian Department of Civil Protection as alert for medium-high seismic events, the frequencies associated with the first two vibration modes of the building have been identified with sufficient accuracy by means of basilar instruments of OMA. Also, the experimental vibration modes of the building associated with the first two frequencies and computed along three instrumented corners have been obtained.

Both experimental frequencies and vibration modes have been compared to the results of linear modal analyses carried out through three refined three-dimensional FE models of the building made of shell elements. The models were aimed to investigate also the role of the subsoil in modifying the dynamic response of the building. In particular, the effect of the subsoil has been introduced by means of concentrated translational and rotational springs placed at the ground floor of the building. The values of stiffness of such springs have been assessed considering well-known literature formulations and based on the mechanical properties of the subsoil.

The comparisons between experimental and numerical results allowed evidencing some flaws both regarding the experimental measures and the FE model. In view of the experimental uncertainty about the identification of higher order modes characterized by a low amount of participating mass and the influence of the type of FE element adopted in the numerical predictions, indeed, only the first two modes of the examined building have been reliably identified.

A good agreement between experimental and numerical results has been finally achieved for the chosen values of elastic properties of the materials and of mass acting on the structure, also taking into account the SSI by means of concentrated springs at the basement.

Even if for the examined building the variation of frequencies and modal shape observed when the SSI is taken into account is not high, a FE model with SSI is surely a better representation of the structure. The examined case study can be considered as representative of a complete approach which accounts also for SSI in the modelling since it allows considering the effect of the subsoil which contributes to the real dynamic behaviour of any building. On the other hand, other types of structure (tower or very deformable buildings) can be even more susceptible to SSI and it is important to model this phenomenon in evaluating their dynamic behaviour.

It is worth pointing out that many parameters may influence the results of modal analyses in terms of both frequencies and modal shapes. In particular, the type of selected finite elements (*i.e.*, a model made of three-dimensional elements is expected to provide higher frequencies compared to a model made of shell elements, since the simulated building is overall more stable), the uncertainty about the geometry of the building, masonry elastic properties (*i.e.*, for the case at hand the values of Young's modulus of masonry should be better assessed with specific *in situ* inquires) and unit weight, the permanent and the accidental loads acting on the building, the mechanical properties of subsoil. Each one of the above factors is worth being singularly examined. This aspect is beyond the scope of this paper, primarily devoted to highlighting the role of the foundation soil in the dynamic identification process of a masonry structure.

The obtained results represent the starting point for future analyses aimed at assessing the seismic safety of the building, located in one of the most seismically active zones of Italy (Sannio).

ACKNOWLEDGMENTS

The Authors would like to thank the Italian Department of Civil Protection (DPC) for sharing the experimental data acquired by the monitoring system.

AUTHOR CONTRIBUTIONS

Contributions of the authors can be estimated in 40% for both Francesca Ceroni and Stefania Sica and 10% for both Angelo Garofano and Marisa Pecce.

CONFLICTS OF INTEREST

The authors declare no conflict of interest.

REFERENCES

1. Brincker, R.; Zhang, L.M.; Andersen, P. Modal identification of output only systems using frequency domain decomposition. *Smart Mater. Struct.* 2001, *10*, 441–445.

2. Gentile, C.; Saisi, A. Ambient vibration testing of historic masonry towers for structural identificatiòn and damage assessment. *Constr. Build. Mater.* 2007, *21*, 1311–1321.

3. Michel, C.; Gueguen, P.; Bard, P.Y. Dynamic parameters of structures extracted from ambient vibration measurements: An aid for the seismic vulnerability assessment of existing buildings in moderate seismic hazard regions. *Soil Dyn. Earthq. Eng.* 2008, *28*, 593–604.

4. Rainieri, C.; Fabbrocino, G. Automated output-only dynamic identification of civil engineering structures. *Mech. Syst. Signal Process.* 2010, *24*, 678–695.

5. Sica, S.; Pagano, L.; Vinale, F. Interpretazione dei segnali sismici registrati sulla diga di Camastra.*Ital. Geotech. J.* 2008, *4*, 97–111, (In Italian).

6. Dhakal, R.P. Explosion-Induced Structural Response: An Overview. In Proceedings of the New Zealand Society of Earthquake Engineering Annual Conference, Rotorua, New Zealand, 19–21 March 2004.

7. Potapov, V.A. Investigation of the explosion-induced vibration of buildings. *Soil Mech. Found. Eng.* 1974, *11*, 170–173.

8. Trifunac, M.D. Comparison between ambient and forced vibration experiments. *Earthq. Eng. Struct. Dyn.* 1972, *1*, 133–150. [Google Scholar]

9. Gueguen, P. Experimental analysis of the seismic response of one base-isolation building according to different levels of shaking: Example of the Martinique earthquake (2007/11/29) Mw 7.3. *Bull. Earthq. Eng.* 2012, *10*, 1285–1298. [Google Scholar]

10. Cimellaro, G.P.; Piantà, S.; de Stefano, A. Output-only modal identification of ancient L'Aquila city hall and civic tower. *J. Struct. Eng.* 2012, *138*, 481–491.

11. Ivorra, S.; Pallarés, F.J. A Masonry Bell-Tower Assessment by Modal Testing. In Proceedings of the 2nd International Operational Modal Analysis Conference (IOMAC), Copenhagen, Denmark, 30 April–2 May 2007.

12. Ceroni, F.; Pecce, M.; Voto, S.; Manfredi, G. Historical, architectural and structural assessment of the Bell Tower of Santa Maria del Carmine. *Int. J. Archit. Herit.* 2009, *3*, 169–194.

13. ICOMOS/ISCARSAH Committee. ICOMOS Charter-Principles for the Analysis, Conservation and Structural Restoration of Architectural Heritage. In Proceedings of the ICOMOS 14th General Assembly and Scientific Symposium, Victoria Falls, Zimbabwe, 27–31 October 2003.

14. De Sortis, A.; Antonacci, E.; Vestroni, F. Dynamic identification of a masonry building using forced vibration test. *Eng. Struct.* 2005, *27*, 155–165.

15. Bennati, S.; Nardini, L.; Salvatore, W. Dynamical behaviour of a masonry medieval tower subjected to bell's action. Part I: Bell's action measurement and modelling. *J. Struct. Eng. ASCE* 2005, *131*, 1647–1655.

16. Hans, S.; Boutin, C.; Ibraim, E.; Rousillon, P. *In situ* experiments and seismic analysis of existing buildings. Part I: Experimental investigations. *Earthq. Eng. Struct. Dyn.* 2005, *34*, 1513–1529.

17. Ditommaso, R.; Parolai, S.; Mucciarelli, M.; Eggert, S.; Sobiesiak, M.; Zschau, J. Monitoring the response and the back-radiated energy of a building subjected to ambient vibration and impulsive action: The Falkenhof Tower (Potsdam, Germany). *Bull. Earthq. Eng.* 2010, *8*, 705–722.

18. Safak, E. Detection and identification of soil-structure interaction in buildings from vibration recordings. *J. Struct. Eng. ASCE* 1995, *121*, 899–906.

19. Laurenzano, G.; Priolo, E.; Gallipoli, M.R.; Mucciarelli, M.; Ponzo, F.C. Effect of vibrating buildings on free-field motion and on adjacent structures: The Bonefro (Italy) case history. *Bull. Seismol. Soc. Am.* 2010, *100*, 802–818.

20. Guéguen, P.; Bard, P.Y. Soil-structure and soil-structure-soil interaction: Experimental evidence. *J. Earthq. Eng.* 2005, *9*, 657–693.

21. Luco, J.E.; Trifunac, M.D.; Wong, H.L. Isolation of soil-structure interaction effects by full-scale forced vibration tests. *Earthq. Eng. Struct. Dyn.* 1988, *116*, 1–21.

22. Stewart, J.P.; Fenves, G.L. System identification for evaluating soil-structure interaction effects in buildings from strong motion recordings. *Earthq. Eng. Struct. Dyn.* 1998, *27*, 869–885.

23. Rainieri, C.; Fabbrocino, G.; Manfredi, G.; Dolce, M. Robust output-only modal identification and monitoring of buildings in the presence of dynamic interactions for rapid post-earthquake emergency management. *Eng. Struct.* 2012, *34*, 436–446.

24. Ceroni, F.; Pecce, M.; Sica, S.; Garofano, A. Assessment of seismic vulnerability of a historical masonry building. *Buildings* 2012, *2*, 332–358.

25. Ceroni, S.; Sica, A.; Garofano, A.; Pecce, M.R. Evaluation of the natural vibration frequencies of a historical masonry building accounting for SSI. *Soil Dyn. Earthq. Eng.* 2014, *64*, 95–101.

26. Zhang, L.; Brincker, R.; Andersen, P. An Overview of Operational Modal Analysis: Major Development and Issues. In Proceedings of the 1st International Operational Modal Analysis Conference (IOMAC), Copenhagen, Denmark, 26–27 April 2005.

27. Trifunac, M.D.; Ivanovic, S.S.; Todorovska, M.I. Apparent periods of a building. I: Fourier analysis. *J. Struct. Eng. ASCE* 2001, *127*, 517–526.

28. Homepage of ISS. Available online: http://www.mot1.it/iss (accessed on 26 November 2014).

29. Veletsos, A.S.; Meek, J.W. Dynamic behaviour of building-foundation systems. *J. Earthq. Eng. Struct. Dyn.* 1974, *3*, 121–138.

30. Gazetas, G. Formulas and charts for impedances of surface and embedded foundations. *J. Geotech. Eng.* 1991, *117*, 1363–1381.

31. Mylonakis, G.; Nikolaou, S.; Gazetas, G. Footings under seismic loading: Analysis and design issues with emphasis on bridge foundations. *Soil Dyn. Earthq. Eng.* 2006, *26*, 824–853.

32. Ceroni, F.; Sica, S.; Pecce, M.; Garofano, A. The Role of the Foundation Soil on the Dynamic Behaviour of Reinforced-Concrete and Masonry Buildings. In Proceedings of the 15th World Conference of Earthquake Engineering (WCEE), Lisbon, Portugal, 24–28 September 2012.

33. Norme Tecniche per le Costruzioni. NTC 2008, Min.LL.PP, DM 14/01/2008. G.U.R.I. No. 29. Available online: http://www.cslp.it/cslp/index.php?option=com_content&task=view&id=66&Itemid=20 (accessed on 24 Novembre 2014). (In Italian).

34. Available online: http://www.cslp.it/cslp/index.php?option=com_content&task=view&id=79&Itemid=1 (accessed on 24 November 2014). (In Italian).

35. Marcari, G.; Fabbrocino, G.; Lourenço, P. Mechanical Properties of Tuff and Calcarenite Stone Masonry Panels under Compression. In Proceedings of the 8th International Masonry Conference, Dresden, Germany, 4–7 July 2010.

36. Sica, S.; Ceroni, F.; Pecce, M.R. Soil Structure Interaction on the Dynamic Behavior of Two Historic Masonry Structures. In Proceedings of the 2nd International Symposium on Geotechnical Engineering for the Preservation of Monuments and Historic Sites, Napoli, Italy, 30–31 May 2013; pp. 657–667.

CITATION

Francesca Ceroni, Stefania Sica, Angelo Garofano and Marisa Pecce. SSI on the Dynamic Behaviour of A Historical Masonry Building: Experimental *Versus* Numerical Results. doi:10.3390/buildings4040978.

CHAPTER 9

The Complex Nature of Pollution in the Capping Soils of Closed Landfills: Case Study In A Mediterranean Setting

Jesús Pastor[1], María Jesús Gutiérrez-Ginés[1], Carmen Bartolomé[2] and Ana Jesús Hernández[2]

[1]Department of Environmental Biology, MNCN, CSIC, Madrid, Spain
[2]Department of Life Sciences, Alcalá University, Alcalá de Henares, Spain

INTRODUCTION

Waste Landfills Capped With Soil

In most developed countries there is serious concern about the state of waste landfills that were closed towards the end of the past century. Economic growth and urban development during this period generated vast amounts of domestic and industrial waste, and this waste was deposited in landfills without its separation or prior treatment. Today, countries with emerging economies or countries in settings of poverty are facing a similar situation, whereby the uncontrolled disposal of waste has led to regions with worryingly high levels of pollutants that affect the atmosphere, soil and water resources.

In the Mediterranean setting, most landfills have been sealed simply by capping with soil from the surroundings. This soil has given rise to a plant cover emerging from the existing seed bank. Besides recovering the visual impacts on the landscape of mountains of rubbish, a plant cover will avoid the spread of pollutants to other ecosystems once the landfill has been closed. However, the implantation of such a cover is conditioned by interactions among several factors, which are responsible for the complex nature of soil pollution affecting closed landfills.

In the literature, there is a lack of work related to pollution of the capping soils of closed landfills. Bibliography concerning landfills is mostly focused on leachates and their effect on water. In case of soil-related works, they were not accomplished in Mediterranean environment. Specially, quantitative data of landfill soil pollution and its possible effects on colonizing plant population is not available.

For all these reasons, the purposes of this chapter are: i) to describe the profile of solid waste landfills sealed with soil in the Mediterranean setting; ii) to focus on the study of a given case in order to present a research methodology that can be used in other scenarios with a similar problem. In the sections detailing this case study, we describe the methods and techniques employed for studying landfill's remediation and discuss the data obtained to give an overview of the topic examined.

PROFILE OF SOLID WASTE LANDFILLS IN THE MEDITERRANEAN REGION

Numerous waste tips in the central Iberian Peninsula capped with soil over the 1980s and 90s have been widely described from an interdisciplinary perspective in [1]. Research efforts have focused on the pollutants present in the soils used to seal 20 of these landfills and on factors inducing the spread of pollutants. These studies have been aimed at designing measures to remediate the visual impacts of solid waste landfills (Figure 1).

Figure 1. Based on [7].

The figure shows the main impacts (not only visual) produced by a landfill sealed with soil. The most important impact is pollution produced by surface and deep leachates of polluting substances generated by surface run off and rainwater infiltration [2-6]. This type of pollution especially affects ecosystems in the main areas of leachates discharge i.e., the foots of landfill slopes grazed by domestic and wild animals [5]. Effects are nevertheless also produced on crops, particularly cereals such as barley, which are sometimes grown on the landfill itself (usually on its platforms). We have noted that this cereal accumulates heavy metals (unpublished data). In addition, soil pollution may spread to nearby rivers, on which these mountains of waste seem to hang, or to streams, which transport pollutants to areas beyond the landfill.

Table 1 summarizes the main plant communities detected at the 20 landfill sites examined, along with a summary of their main characteristics according to [8]. Despite the 20 years passed since the landfills were closed, plant cover generally lacks a bush stratum. Existing communities are those classified as ruderals and nitrophiles with a dominance of annual species whose life cycle is typical of the Mediterranean region. Many of these landfills still show large expanses of soil unable to sustain plant growth while other areas boast good plant cover, though with a low diversity. In general, all landfill sites are grazed by itinerant herds of sheep.

Table 1. Phytosociological classes and mean characteristics of the main species found at the landfills

Phytosociological class	Main characteristics
9. ISOETO-NANOJUNCETEA Br.-Bl. & Tüxen ex Westhoff, Dijk & Passchier 1946	Pionner annual and dwarf perennial ephemeral isoetid communities on periodically flooded bare soils
12. PHRAGMITO-MAGNOCARICETAE Klika in Klika & V. Novák 1941	Swampy, fenny, lacustrine and riverine helophyte communities dominated by perennial graminoids, sedges, forbs and herbs of fresh and brackish waters
20. JUNCETEA MARITIMI Br.-Bl. in Br.-Bl. Roussine & Nègre 1952	Perennial grasslands growing on coastal and inland temporary wet or inundated salt marshes
34. ARTEMISIETEA VULGARIS Lohmeyer, Preising & Tuxen ex von Rochow 1951	Perennial and tall biennial forbs, grasses and thistle pioneer ruderal and nitrophilous sunny communities growing on rich soils.
37. PEGANO-SALSOLETEA Br.-Bl. & O. Bolós 1958	Nitrophilous or halo-nitrophilous dwarf scrub communities, including anthropogenic alloctonous shrubby vegetation
38. POLYGONO-POETEA ANNUAE Rivas Martinez 1975	Annual pioneer ephemeral and exceptionally small creeping perennial nitrophilous anthropozoogenic heavy trodden communities of urban and rural paths
39. STELLARIETEA MEDIAE Tüxen, Lohmeyer & Preising ex von Rochow 1951.	Annual ephemeral weeds, ruderal, nitrophilous and semi-nitrophilous communities
39A. Stellarienea mediae	-Cultivated field weed communities
39B. Chenopodio-Stellarienea Rivas Goday 1956	-Ruderal, nitrophilous and seminitrophilous communities.
39e. Thero-Brometalia (Rivas Goday&Rivas Martinez ex Esteve 1973) O. Bolós 1975	-Subnitrophilous Mediterranean annual ephemeral grassland-like spring blooming communities.

39f. Sisymbrietalia officinalis J. Tüxen in Lohmeyer & al. 1962 em. Rivas-Martinez, Fernándz-González & Loidi 1991

- Nitrophilous and temperate annual ephemeral grassland-like. Path, roadside and rural often trampled communities

40. GALIO-URTICETEA Passarge ex Kopecký

Perennial hemycriptophyte and climbing tall herbs of nitrified wood fringes and other semi-shaded anthropogenic biotope communities.

41. CARDAMINO HIRSUTAE-GERANIETEA PURPUREI (Rivas-Martínez, Fernández-González & Loidi 1999) Rivas-Martínez, Fernández-González & Loidi classis nova, stat. Nov.

Annual spring and summer ephemeral internal and external shrub fringes slightly nitrified semi-shaded communities, growing on rich organic nutrient soils.

43. TRIFOLIO-GERANIETEA Müller 1962

Semi-shaded perennial herb communities of scarce moisture external fringe woodlands. Calcareous or mesoeutrophic rich soils in temperate submediterranean central Iberian territories.

50. TUBERARIETEA GUTTATAE (Br.-Bl. in Br.-Bl., Roussine & Nègre 1952) Rivas Goday & Rivas Martínez 1963 nom. mut. Propos.

Therophytic grasslands. Pioneer spring and early summer ephemeral plant acidophilous or calcifugous communities, dominated by non nitrophilous annual short herbs and grasses, but localized only in dry or initial soils, mostly in submediterranean or step territories.

51. FESTUCO-BROMETEA Br.-Bl. & Tüxen ex Br.-Bl. 1949

Perennial xerophytic and mesophytic grasslands. Anthropogenic grazed baso-neutrophilous or slightly acidophilous mesophytic or slightly xerophytic nutrient rich-pastures largely covered by perennial grasses.

54. POETEA BULBOSAE Rivas Goday & Rivas Martínez in Rivas-Martinez 1978

Western Mediterranean oceanic thermo- to supramediterranean upper semiarid to humid pastures, grazed and manured, dominated by dwarf perennial grasses and other nutritious prostrate chamaephytes.....

56. LYGEO-STIPETEA Rivas-Martinez 1978 nom. Conserv. Propos.

Mediterranean perennial basophilous xerophytic tall bunchy dense or short open grasslands.

57. STIPO GIGANTAE-AGROSTIETEA CASTELLANAE Rivas-Martínez, Fernández-González & Loidi 1999	Silicicolous perennial grasslands rich in endemics, serial of *Quercus rotundifolia* and other *Quercus* natural potential forest communities.
59. MOLINIO-ARRHENATHERETEA Tüxen 1937	Mesophile to wet often manured meadows and pasture communities on deep and moist soils, widely spread by grazed and anthropic activities

Most of the capped landfills are mixed dumps containing both domestic and industrial waste. Besides mitigating the visual impacts of a landfill, the plant cover prevents its collapse and the pollution of other ecosystems by deposited waste materials.

However, in such scenarios the stability of plant communities that become established from the seed bank of the capping soil layer is threatened. Among others, the factors that give rise to this situation are continued waste disposal after the landfill's initial sealing, the scarce volume of capping soil present and land use projects implemented without a priori planning.

CASE STUDY: THE GETAFE LANDFILL (MADRID)

Geomorphological Characterization

Here we examine the case of a closed landfill in the Madrid Autonomous Community. This site can be described as one of the most complex scenarios observed among the soil-capped solid waste landfills of the central Iberian Peninsula despite its many features common to all the landfills examined in this region [1]. Located in the municipal district of Getafe (Madrid), this landfill was first described by [9], when it occupied an area of around 70,000 m³. Fifteen years later (in 2009), the site covered some 95,000 m² of land.

Continuous waste dumping and subsequent capping with soil from the surroundings has determined the complex morphology of this landfill. In the photo in Figure 2, the landfill appears as a flattened hill rising out of a plain.

Figure 2. Picture of the whole landfill in spring of 2009.

The landfill site has three main zones: a zone (western) mostly containing solid domestic waste, and two zones (central and eastern) mainly accommodating industrial waste and some inert compounds. We have designated these latter zones "rubble tips" to distinguish them from the landfill proper (Figure 3).

The flatted tops of the landfill correspond to platforms, yet more outstanding are its 12 slopes showing a high variety of exposures (across their 360°). Slope heights are 10-20 m and gradients are 50%. Their profiles are straight and many slopes overlap one another. Many slopes show signs of erosion, especially in troughs, often exposing their waste materials. Leachate surface runoff may be observed in three main discharge areas. The westernmost discharge area occupies a wetland. The other two areas, south of the rubble tips form shallow water sheets in the wettest months and quickly dry when rain ceases at the end of spring. In all these discharge zones and at the foots of the slopes, sheep herds may be found grazing. What is more, these and other animals drink any water that accumulates in these areas in 5 to 6 months of the year.

Composition of the Capping Soil Layer: Factors Linked to Fertility, Salinity, Metal Toxicity, Organic Compounds and Erosion

To identify the soil factors that mainly determine the landfill's vegetation, mostly arising from the seed bank of the capping soil, we used a stratified sampling procedure (platforms, adjacent rubble tips and main surface leachate discharge zones). At each site, samples were collected using a hoe from the top soil layer (0-10 cm) to give an average soil sample. 57 of such samples were transported to the laboratory, where they were air-dried and sieved (< 2 mm). These samples were then subjected to each of the techniques mentioned in the following sections. Sampling sites 27 to 31 correspond to piles of waste deposited directly in the easternmost discharge zone with no type of cover at all. Although these samples do not correspond to the capping soil, they were collected to assess the possible effect of these waste materials on the soils of the discharge zone in future studies. Corresponding results do not appear in the tables provided below.

Figure 3. Main areas of the landfill and sites where capping soil samples were collected. 03/08/2009 Google Earth Image. UTM coordinates: X=442599 Y=4459459, 30T.

Soil Fertility Indicators

In the 57 soil samples, we examined several variables related to soil fertility with consequent impacts on the vegetation. The procedures described in [10] were used to determine: pH in water and in a saturated soil paste, percentage organic matter by potassium dichromate reduction, Kjedahl total nitrogen, pseudototal (by extraction with nitric and perchloric acids at 4:1) and exchangeable (by extraction with ammonium acetate, pH 7) concentrations of Ca, K, Na and Mg, and pseudototal and bioavailable P and P_2O_5 concentrations, analyzed by plasma emission spectroscopy (ICP-OES).

Table 2. pH, organic matter (OM, %), nitrogen (N, %), pseudo-total concentration (T) of nutrient elements (mg kg^{-1}) and percentage of exchangeable fraction (E) in soil samples collected from landfill proper

Sampling point	pH	O.M.	N	Ca		K		Na		Mg		P		P_2O_5
				T	E	T	E	T	E	T	E	T	E	
Platform														
G-23	8.5	2.9	0.09	61566	6	1729	5	923	2	7111	3	205	0	130
G-24	8.0	1.2	0.06	5483	6	8855	4	244	5	4869	6	598	7	240
G-45	8.1	1.3	0.08	7345	3	6892	4	223	4	7085	4	553	3	100
G-46	7.8	1.4	0.08	15253	28	11362	7	276	6	22144	4	542	8	260
G-47	8.3	1.8	0.10	14890	22	10051	3	257	5	11553	4	569	4	110
G-48	7.9	1.5	0.10	9818	34	9400	5	218	7	11036	5	462	7	150
Slope														
G-6	7.5	4.0	0.05	6627	27	2495	5	86	16	1316	7	70	0	40
G-8	8.1	0.9	0.05	42871	9	3600	7	355	6	10842	2	32	0	50
G-11	8.2	2.5	0.12	42156	12	6790	12	337	16	7992	4	542	4	300
G-16	7.8	2.2	0.12	46571	14	4478	14	502	12	5795	5	317	0	110
G-17	8.0	2.3	0.06	57028	8	2524	10	604	6	6123	6	132	0	45
G-18	8.2	2.8	0.18	38558	9	3262	9	760	15	6124	7	113	0	15
G-19	7.6	1.2	0.19	11643	27	8270	4	241	15	6617	8	441	0	60

G-20	7.8	1.4	0.09	4051	59	11010	5	243	6	6765	9	547	9	240
G-21	7.9	0.6	0.05	8366	41	7671	4	222	7	5506	8	494	8	200
G-22	8.1	0.9	0.06	7296	53	9209	4	254	5	5810	5	590	6	190
G-39	8.1	1.1	0.05	25150	8	4800	4	350	6	8250	3	50	0	20
Foot of slope													
G- 3 A	7.9	0.9	0.06	3969	53	6902	7	221	27	6222	7	258	10	140
G- 3 F	7.6	2.1	0.09	5922	46	7492	6	214	18	10126	5	258	14	200
G-9	8.2	1.2	0.06	42176	10	3206	7	467	5	4391	4	106	0	60
G-10	8.2	0.9	0.03	29893	9	4759	4	339	5	5086	5	137	0	20
G-37	7.8	3.7	0.17	10122	41	8360	8	224	9	4598	7	429	19	320
G-38	7.8	2.3	0.11	14063	35	5774	5	215	32	4861	11	444	9	200
Discharge zone													
G- 2 A	7.7	6.2	0.29	13490	39	7245	10	324	52	9071	6	279	9	140
G- 2 B	7.8	7.8	0.37	28560	21	6702	9	252	34	8336	7	366	12	220
G- 4 A	7.5	3.8	0.13	12337	20	3570	12	162	26	3884	5	10	0	1050
G- 4 B	7.8	3.8	0.10	11135	19	3468	8	177	32	3577	8	8	0	70
G-7	7.5	5.6	0.25	37831	33	7442	3	308	32	7953	10	276	0	115

The results of these determinations in each soil sample are provided in Tables 2 and 3. All results are provided to highlight the huge variation existing for each factor. pH varied from 7.0 to 8.5, given the alkaline nature of the surrounding soils used to cap the landfill. The distributions of all variables failed to vary significantly between the landfill proper and rubble tips.

Heavy Metals and Trace Elements Toxic for Plants

Although some preliminary results regarding soil pollution due to heavy metals, organic compounds and salinity have been already described [1, 7, 11, 12] here we examine this issue in detail. All 57 soil samples were subjected to inductively coupled plasma optical emission spectometry (ICP-OES) to determine pseudototal (after prior extraction with nitric and perchloric acids, 4:1 [13]) and bioavailable (after prior extraction with ammonium acetate + EDTA using the method [14]) concentrations of Al, Mn, Zn, Cu, Pb, Cr and Ni. In addition, total As concentrations were determined by X-ray fluorescence in 48 samples. Total Hg levels were determined using an Advanced Mercury Analyser (AMA-254, LECO Company, Czeck Republic) according to the procedure described by [15] in 34 selected samples of the 57 soil samples collected [16].

Table 3. pH, organic matter (OM, %), nitrogen (N, %), pseudo-total concentration (T) of nutrient elements (mg kg^{-1}) and percentage of exchangeable fraction (E) in soil samples collected from rubble tips.

Sampling point	pH	O.M.	N	Ca		K		Na		Mg		P		P$_2$O$_5$
				T	E	T	E	T	E	T	E	T	E	
Platform														
G- 49	8.1	1.7	0.11	13588	24	7139	7	102	15	10111	5	613	3	18
G- 50	7.6	1.6	0.12	10910	38	7464	7	120	11	9279	4	297	11	1171
G- 51	7.8	2.2	0.10	14428	23	9093	5	134	8	10778	2	873	6	66
G- 52	7.8	0.8	0.04	5342	42	6711	3	106	11	6411	4	283	4	12
G- 53	7.7	2.3	0.10	15831	21	7671	8	97	12	7853	3	580	12	54
Slope														
G- 13	8.3	1.5	0.08	26245	16	2653	16	187	11	3343	6	106	21	30
G- 15	7.7	1.2	0.09	16589	50	7331	5	198	8	4009	7	249	4	8603
G- 32	7.8	0.7	0.05	63966	14	6989	33	238	12	7583	5	261	3	4008
G- 40	8.1	1.0	0.04	14200	28	7550	3	250	6	10350	2	350	4	264

G- 42	8.2	0.2	0.03	9617	35	3704	4	158	16	4643	4	289	12	26
Foot of slope														
G- 12	7.9	1.4	0.07	10790	37	7968	4	214	7	7278	9	423	4	144
G- 14	8.0	1.0	0.06	17651	27	3658	7	127	11	3126	9	68	0	159
G- 33	7.9	1.4	0.09	17519	27	7125	6	190	7	5067	6	536	6	72
G- 34	7.0	2.5	0.13	31469	17	7942	5	248	22	6187	7	469	5	790
G- 35	7.9	1.2	0.08	25913	19	8208	5	173	12	7176	7	428	5	44
G- 41	7.9	1.1	0.10	24281	20	8442	4	246	6	8838	1	593	19	25903
G- 43	8.0	0.8	0.05	72211	6	7503	3	246	6	10099	3	362	8	993
G- 44	7.9	1.3	0.07	17827	19	7959	3	315	13	5894	5	253	4	287
Discharge zone														
G- 1 A	7.7	4.2	0.20	72005	34	4234	5	1281	74	8466	22	177	10	5918
G- 1 B	7.7	2.0	0.11	43501	21	7456	4	292	19	7316	3	265	6	3830
G- 5	7.3	5.8	0.24	25041	30	8150	8	412	40	11338	3	449	6	3305
G- 25	8.0	2.6	0.12	15164	28	8552	6	310	27	5753	10	407	5	74
G- 26	7.8	1.1	0.07	64650	9	6624	3	198	15	6081	5	240	4	819
G- 36	7.5	5.6	0.30	34429	25	9961	5	306	27	8024	8	498	3	4245

Tables 4 and 5 provide the metal and trace element concentrations detected in the capping soil and discharge area samples. We also examined total Al and Mn levels: Al concentrations in the landfill ranged from 8123 to 50747 mg kg^{-1}, and Mn concentrations from 205 to 7432 mg kg^{-1}. In the rubble tips, concentration ranges were higher for Al and lower for Mn. Given the alkaline nature of the soils, these elements are not considered hazardous for plant populations and are therefore not included in the tables.

Table 4. Pseudo-total concentration (T) of trace elements (mg kg^{-1}) and percentage of bioavailable fraction (B) in soil samples collected from landfill proper. nd: not detected; -: not analyzed. Reference levels for alkaline soils according to Spanish law (RD1310/1990), *As Dutch reference level.

Sampling point	Zn		Cu		Pb		Cd		Cr		Ni		As	Hg
	T	B	T	B	T	B	T	B	T	B	T	B		
Platform														
G-23	9491	23	4593	3	4421	50	93	35	531	0.9	231	1.1	271	11
G-24	137	20	14	18	19	70	0.0		3.2	2.2	8.7	1.5	n.d	0.0
G-45	148	2.1	8.9	13	28	8.2	0.0		7.3	0.6	5.6	4.1	14	-
G-46	147	2.1	38	17	35	48	0.0		2.2	2.3	12	4.0	22	-
G-47	126	3.8	12	14	11	43	0.0		4.7	1.1	9.9	3.4	n.d	-
G-48	175	5.3	43	16	22	40	0.0		18	0.6	21	3.4	n.d	-
Slope														
G-6	640	22	1916	21	132	29	0.0		256	0.1	35	2.5	119	0.3
G-8	13029	27	1055	8.1	12689	23	185	34	269	1.7	86	1.9	282	4.0
G-11	3247	20	240	11	579	51	10	67	154	0.6	43	2.0	n.d	0.5
G-16	10777	25	748	9.3	5734	38	142	41	150	3.0	42	2.6	n.d	3.0
G-17	17416	26	1313	8.5	12612	28	308	37	242	1.5	60	2.2	n.d	4.9
G-18	22992	28	1804	9.4	18136	30	306	36	298	1.8	80	3.7	685	4.2
G-19	5190	25	125	14	1198	47	29	52	40	3.3	15	2.6	n.d	0.2
G-20	1085	2.6	18	9.4	106	12	0.3	64	9.8	0.5	9.0	0.9	12	0.0
G-21	129	9.0	11	21	18	37	0.0		3.5	1.4	7.0	1.7	13	0.0
G-22	168	9.3	14	20	25	53	0.0		9.1	0.8	8.8	1.8	23	0.0
G-39	17830	76	1445	20	12555	68	190	60	587	2.2	210	2.5	492	-
Foot of														

slope														
G- 3 A	61	2.2	9.1	14	18	59	0.0		2.3	12	5.1	9.2	-	-
G- 3 F	69	4.6	31	6.2	8.1	40	0.0		9.2	3	5.2	7.0	-	-
G-9	12632	25	1181	10	7179	37	257	48	236	2.4	90	2.1	306	2.4
G-10	15184	33	1493	13	10085	48	155	42	504	2.2	206	2.2	n.d	3.3
G-37	559	66	20	25	28	61	0.0		3.8	4.7	8.5	5.9	15	0.0
G-38	108	18	15	24	23	65	0.0		15	1.0	7.4	6.7	19	0.0
Discharge zones														
G- 2 A	528	17	36	27	117	36	0.0		6.6	6.1	8.6	5.1	-	-
G- 2 F	366	14	35	18	64	28	0.0		5.7	4.9	7.3	5.5	-	-
G- 4 A	577	19	882	27	149	29	0.0		110	0.1	156	1.5	-	-
G- 4 F	477	19	1260	26	139	19	0.0		153	0.2	213	1.5	-	-
G-7	2417	33	151	27	290	48	2.5	86	9.5	3.4	34	8.9	54	0.3
Ref. (pH"/> 7)	450		210		300		3		150		112		29*	1.5

The sites showing the highest levels of all elements occurred on the landfill's slopes and these showed an uneven spatial distribution. However, the most contaminated sites were simultaneously polluted by all elements. The percentage of a metal found in its bioavailable form was also highly variable. Despite being poorly mobile, Pb showed high bioavailability percentages. Cd was also highly bioavailable. Most variation was shown by Zn and Cu. The metals appearing in lowest concentrations were Cr and Ni.

Apart from the trace element bioavailability study conducted according to the method of [14], we performed a more exhaustive analysis of metal bioavailability in the soil samples. To this end, we undertook sequential extraction by the BCR method optimized by [17]. Sequential extraction serves to indicate the fractions of each metal that are bioavailable (F1: exchangeable), reducible (F2: bound to oxyhydroxides), oxidizable (F3: bound to organic matter) and residual (F4). Given that it is the landfill proper that shows the higher concentrations of these metal pollutants, 5 sites were selected representing platform, slope and discharge zones showing variable concentrations of these types of pollutant. The sites were selected according to their known distributions of metals; we have preserved the numbers assigned to their collection sites. Table 6 provides total concentrations of each metal in each sample calculated as the sum of all fractions. The reader may find the percentages of each metal found in each fraction in Figure 4.

Table 5. Pseudo-total concentration (T) of trace elements (mg kg^{-1}) and percentage of bioavailable fraction (B) in soil samples collected from rubble tips. nd: not detected; -: not analyzed. Reference levels for alkaline soils according to Spanish law [18], *As Dutch reference level.

Sampling point	Zn		Cu		Pb		Cd		Cr		Ni		As	Hg
	T	B	T	B	T	B	T	B	T	B	T	B		
Platform														
G-49	10	0.0	13	21	0.0		0.0		1.8	0.0	5.4	0.0	n.d	-
G-50	17	0.0	21	23	22	54	0.0		3.4	0.0	7.8	0.0	n.d	-
G-51	13	0.0	17	19	0.0		0.0		2.4	0.0	6.5	0.0	n.d	-
G-52	9	0.0	14	18	0.0		0.0		3.6	0.0	8.0	0.0	12	-
G-53	13	0.0	14	29	0.0		0.0		6.8	0.0	7.7	0.0	15	-
Slope														
G-13	1261	18	783	7.0	164	35	0.0		378	0.1	223	1.2	42	0.8
G-15	159	15	34	14	28	48	0.0		17	0.7	12	3.3	29	0.0
G-32	166	11	40	54	24	42	0.0		14	3.8	11	3.4	26	0.1
G-40	120	35	15	14	15	86	0.0		3.5	3.4	5.0	6.6	14	-
G-42	72	12	10	9.2	10	43	0.0		6.7	1.5	4.1	7.1	n.d	-

Foot of slope														
G-12	1123	2	83	14	129	10	0.0		10	1.2	25	2.6	25	0.0
G-14	341	16	266	11	64	32	0.0		94	0.1	75	1.6	22	0.2
G-33	146	19	29	14	26	56	1.2	90	7	2.5	8.8	3.5	22	0.0
G-34	224	18	28	22	48	40	1.5	3.6	10	1.9	12	4.4	37	0.1
G-35	76	4.8	14	18	11	33	0.9	24	13	1.2	9.1	3.2	31	0.0
G-41	1259	1.3	38	13	148	7.6	1.4	26	9.1	1.2	11	4.1	21	-
G-43	226	2.8	22	5.4	24	12	0.0		1.1	9.1	10	2.2	100	-
G-44	1869	10	73	21	215	54	3.5	67	18	0.9	22	3.6	30	-
Discharge zone														
G- 1 A	238	23	18	32	57	43	0.0		2.9	12	4.6	9.0	-	-
G- 1 F	74	4.0	13	19	17	22	0.0		1.5	18	6.4	5.2	-	-
G-5	252	65	56	31	66	41	0.0		10	0.1	8.3	5.2	31	0.2
G-25	171	13	25	19	44	45	0.0		5.7	3.3	10	4.0	27	0.4
G-26	126	8.1	63	8.5	83	32	0.0		24	1.0	16	2.0	36	0.2
G-36	904	35	68	41	152	62	1.7	11	9.2	2.8	14	7.5	40	0.6
Ref. (pH"/>7)	450		210		300		3		150		112		29*	1.5

The results of these tests prompt the following conclusions. Cd and Zn showed highest percentages in the bioavailable fraction. In the case of Cd, this finding is of major concern given its high concentration in the soils examined and this situation has been also described by [19]. Arsenic shows the highest residual percentage and thus its available levels are low. The bioavailable fraction of Cu is fairly low, while remaining fractions vary according to the soils. The behavior of Pb was more irregular among the different soils. In general, the residual fraction was low. The organic matter fraction was variable being greatest at site 18. Its bioavailable fraction was very high at site 39, which is worrying given the high concentration of this heavy metal at this site.

Table 6. Total concentration of trace elements (mg kg^{-1}) in soils selected for conducting the sequential fractionation.

Sample		Zn	Cu	Pb	Cd	As
G-7	wetland	1425	102	278	3	16
G-18	slope	127529	1441	19500	364	32
G-23	platform	33446	2042	11057	106	10
G-39	slope	76120	1272	20032	223	14
G-46	platform	610	41	182	2	0

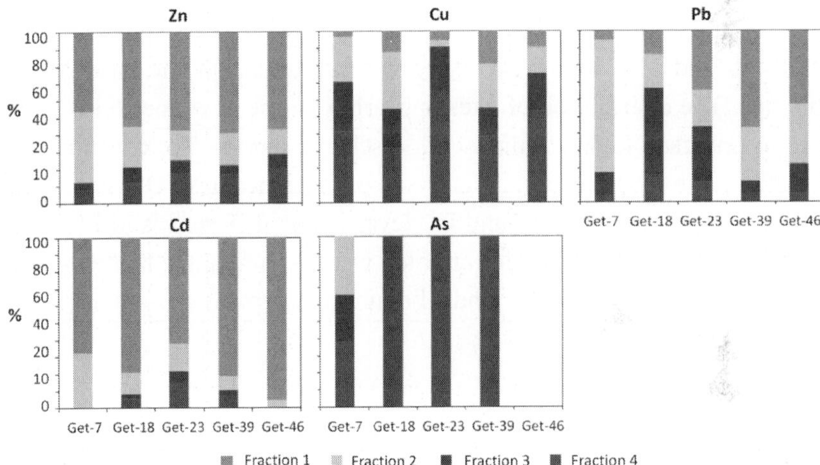

Figure 4. Percentage of metal content that is in each fraction.

Clearly the presence of very high concentrations of trace elements and heavy metals is a problem for the establishment of plant populations, worsened by the fact that the sites of highest concentrations coincide with zones of intense slope. In effect, zones corresponding to samples 17 and 18, along with 10 and 39, are naked slopes with practically no plant cover in comparison with surrounding zones.

Salinity

Salinity has been described as one of the main impacts on the plant populations and animals of the closed landfills of the Iberian Peninsula's central region [5, 12, 20, 21] as well as landfills in other environments [22-25]. This problem is therefore closely examined in the case of the Getafe landfill. Electrical conductivity was determined in all the soil samples collected, along with F^-, Cl^-, NO_2^-, NO_3^-, PO_4^3 and SO_4^{2-} anion concentrations by ion chromatography. These results may be found in Tables 7 and 8. As for the other chemical properties of the soil, points showing greatest salinity were unevenly distributed throughout the capping soil.

By comparing soils from the landfill and rubble tips by principal components analysis (PCA) (Statgraphics 15), we were able to observe that electrical conductivity was related to the Cl^- and SO_4^{2-} contents of the landfill cap and also to F^-, Na, NO_3^- concentrations in the case of the rubble tips. The distribution of sites appearing on the new coordinate axes seems to indicate higher salinity in discharge areas. To confirm this observation, we conducted an analysis of variance (ANOVA) of the factors electrical conductivity and Cl^- level in the different landfill areas. Our findings indicate that both variables were significantly higher in the soils of the wetlands where the landfill's runoff is deposited.

Table 7. Electrical conductivity (EC, $\mu S\ cm^{-1}$) and anion concentration (mg kg^{-1}) in soil samples collected from landfill proper.

Landfill samples	EC	F^-	Cl^-	NO_2^-	NO_3^-	PO_4^{3-}	SO_4^{2-}
Platform							
G- 23	132	5.5	4.6	1.4	5.4	0	26
G- 24	107	1.8	8.1	1.6	52	2.1	11
G- 45	114	2.6	12	1.5	3.3	2.1	8.5
G- 46	161	7.4	8.5	1.1	5.3	2.3	31
G- 47	116	2.4	8.9	1.7	7.6	1.6	10
G- 48	113	2.6	13	1.6	3.7	3.1	8
Slope							
G- 6	175	2.8	6.4	0.7	3	0	147
G- 8	317	19	6	1.4	17	0	398

G- 11	159	7.6	7.7	1	3.6	0	41
G- 16	1853	10	38	1.3	70	0	4063
G- 17	587	14	10	1.4	22	0	592
G- 18	985	31	121	2.6	193	0	635
G- 19	1329	13	32	2	###	0	691
G- 20	154	1.6	5.2	1.3	38	2.2	15
G- 21	171	2.5	6.3	0.9	94	0.9	91
G- 22	157	1.7	4.9	1.1	15	1.1	123
G- 39	366	18	8.6	3	41	0	214
Foot of slope							
G- 3 A	270	3.2	33	1.2	28	0	96
G- 3 F	340	1.6	30	1.5	110	5.4	67
G- 9	212	18	8.7	1.3	13	0	139
G- 10	237	17	5.3	1.1	12	0	194
G- 37	391	1.5	27	1	112	0	65
G- 38	369	4.2	60	1.3	7.3	0	656
Discharge zones							
G- 2 A	1490	3.8	135	2.9	93	2	1938
G- 2 F	1960	1.6	77	1.5	0	0	3322
G- 4 A	1490	3.7	125	1.9	148	2.9	1647
G- 4 F	1500	2.9	43	1	34	0	1729
G- 7	1878	10	85	1.3	20	1	4866

Organic Compounds

Pollution by organic compounds is also a concern emerging from studies designed to address the topic of sealed landfills, as many recently banned compounds, dumped in landfills and numerous affected ecosystems, have been detected [26]. The organic compounds determined in the soil samples and the techniques used for this purpose were: total hydrocarbons by infrared spectrometry (UNE 77307); organochlorine insecticides and polychlorinated biphenyls (PCBs) by gas chromatography (ISO 10382); and polycyclic aromatic hydrocarbons (PAHs) (ISO 18287) and phenols (U.S. E.P.A 3550B, U.S. E.P.A 3650B and U.S. E.P.A 8401) by gas chromatography. The reader is referred to [1] for descriptions of these techniques and their modifications for the present purposes.

Table 8. Electrical conductivity (EC, $\mu S\ cm^{-1}$) and anion concentration (mg kg^{-1}) in soil samples collected from rubble tips.

Tip samples	EC	F^-	Cl^-	NO_2^-	NO_3^-	PO_4^{3-}	SO_4^{2-}
Platform							
G-49	135	1.5	6.6	0	0	5.2	18
G-50	826	1.1	3.2	2.4	0	0	1171
G-51	169	1.7	8.3	0	0	5	66
G-52	60	2.4	2.9	2.6	0	2.8	12
G-53	206	1.2	5.1	2.6	0.6	8	54
Slope							
G- 13	170	7.3	7.2	0.8	18	0	30
G- 15	1719	3.7	4.4	0.8	12	0	8603
G- 32	2360	9.4	6.9	1	67	0	4008
G- 40	299	2.8	5.1	1	3.8	0	264
G- 42	99	3	7.9	1.5	4.3	1.8	26
Foot of slope							
G- 12	90	0	1.1	0	0	0	144
G- 14	193	4.2	15	1.9	13	0	159
G- 33	327	2.5	13	1	143	0	72
G- 34	744	3.8	56	1.5	21	1.5	790
G- 35	198	2.7	13	2.1	37	2.8	44
G- 41	1716	1.9	4.8	1.2	19	0	25903
G- 43	877	2.1	3.6	1	11	0	993
G- 44	384	4.4	46	1.4	5	0	287
Discharge zones							
G- 1 A	8220	8.6	7570	0	495	0	5918
G- 1 F	2350	2.9	199	2.3	29	0	3830
G- 5	2180	3.8	260	2	201	0	3305
G- 25	280	3.3	35	1.9	24	0	74
G- 26	709	6	12	1.5	60	0	819
G- 36	2500	5.2	69	3.6	43	0	4245

Table 9. Maximum concentration of organic pollutants found in soils of Getafe landfill (mg kg^{-1}) and maximum allowed values according to Spanish law (*Ref, [27]).

Pollutant	Max conc.	Ref*	Pollutant	Max conc.	Ref*
PAHs			*Insecticides*		
Naphthalene	0.23	1	Alfa-HCH	0.01	0.01
Acenaphthene	0.04	6	Beta-HCH	0.27	0.01
Fluorene	0.09	5	Gamma-HCH	0.48	0.01
Anthracene	0.46	45	Hexachlorobenzene	0.04	0.01
Fluoranthene	2.59	8	Endosulfan	0.07	0.6
Pyrene	2.03	6	p.p'-DDE	0.02	0.6
1,2-benzanthracene	0.99	0.2	*Hydrocarbons*		
Chrysene	1.11	20	Total concentration	3408	50
Benzo(b)fluoranthene	1.85	0.2	Conc. of aromatics	335	
Benzo(k)fluoranthene	0.89	2	Conc. of aliphatics	3073	
Benzo(a)pyrene	1.56	0.02	*Phenols*		
Indene-1,2,3-(cd)pyrene	1.63	0.3	Phenol	0.05	7
Dibenzo(a,h)anthracene	0.14	0.03	Cresols	0.02	4
			2,4,6-trichlorophenol	0.01	0.9
PCBs	3.05	0.01	Pentachlorophenol	0.01	0.01

Table 9 shows the great variety of organic pollutants that may be found in the Getafe landfill. Those detected at concentrations higher than permitted levels and widely distributed at the site were total hydrocarbons, PCBs, the PAHs with a greater number of rings and some organochlorine insecticides. In general, the sites showing most pollution of this type were those also showing most heavy metal pollution.

Given that total hydrocarbons were detected at all the sites in which these were examined (N=43),Table 10 presents the differences detected.

Table 10. Total concentration of hydrocarbons (HC, mg kg^{-1}) in points of landfill proper and rubble tips and maximum allowed values according to Spanish law (Ref, [27])

Landfill		Landfill		Tips	
Sampling point	HC	Sampling point	HC	Sampling point	HC
Platform		*Foot of slope*		*Slope*	
G-23	901	G- 3 A	13	G-13	215
G-24	78	G- 3 F	13	G-15	52
Slope		G-9	92	G-32	33
G-6	2423	G-10	154	*Foot of slope*	
G-8	93	G-37	33	G-12	174
G-11	230	G-38	26	G-14	169
G-16	62	*Discharge zone*		G-33	52
G-17	63	G- 2 A	5.1	G-34	19
G-18	3408	G- 2 F	5.1	G-35	22
G-19	95	G- 4 A	854	*Discharge zone*	
G-20	33	G- 4 F	854	G- 1 A	7.5
G-21	42	G-7	123	G- 1 F	7.5
G-22	78			G-5	67
G-39	87			G-25	24
				G-26	18
				G-36	36
Ref.	50	Ref.	50	Ref.	50

Factors linked to soil erosion

Signs of soil erosion observed on the landfill's slopes prompted us to address this matter, given the significant effect that soil particle size and the loss of certain fractions can have on the ability of plant species to take root.

The traditional method of Bouyoucos to determine sand, mud and clay fractions was used on all 57 soil samples. In addition, the Mastersizer-S was used to assess particle size by the dispersion and diffraction of a laser light beam as it crosses a suspension of the sample. This technique and the sample preparation method are described in [1]. Particle size was determined in 43 of the samples to establish the type of particle that may be lost through erosion. Significant differences in this variable were detected in several fractions of fine sand between soil from the landfill cap and soil from the rubble tips. These results are provided in Figure 5 and table 11. The high standard deviation of the data determined that only differences in the sand fraction of the rubble tip soil were significant.

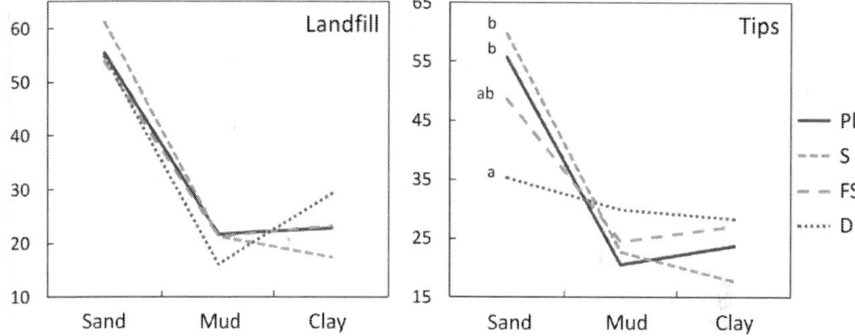

Figure 5. Mean percentage of each textural fraction determined by Bouyoucos technique in samples from platforms (P), slopes (S), foots of slopes (FS) and discharge zones (D) in landfill proper and tips. Different letters mean significant differences between means in the same area (Bonferroni, 95%).

Although the results obtained using both granulometric techniques are not comparable since the first method gives a percentage weight while the second procedure provides percentage volumes, both revealed that the most marked differences among the higher zones, slopes and lower zones occur in the rubble tips adjacent to the landfill. Table 11 shows the different granulometric fractions analyzed. For the rubble tips, results indicate the dragging of fine sands from slopes towards the lower zones accompanied by the consequent build-up of coarse sands. Although with a lack of significance, differences were also observed in the remaining fractions.

These data do not seem to clearly indicate the signs produced by the in situ transport of particles from the higher to the lower zones of slopes and discharge areas. No distinguishing factors were revealed in a discriminatory analysis (figure 6). The findings of such a study also indicate the heterogeneity of the situations arising on even a single slope and increase the complexity of understanding the plant colonization pattern, which may vary as small patches depending on these variations produced on a small scale.

Table 11. Mean (M) and standard deviation (SD) of percentages of each granulometric fraction in different areas of landfill and tips. Different letters in the same range of particle size mean significant differences between means (Bonferroni, 95%).

Area		Range of particle size (mm)							
		Clay	Mud	Fine sand A	Fine sand B	Fine sand C	Medium sand	Coarse sand	Very coarse sand
		<0.002	0.002-0.02	0.02-0.05	0.05-0.1	0.1-0.2	0.2-0.5	0.5-1	1-3.2
Landfill									
Platform	M	0.34	1.93	2.69	7.64	13.7	16.4	23.8	33.4
	SD	0.23	0.65	1.39	4.28	6.72	3.1	9.05	14.5
Slope	M	0.46	3.12	3.32	8.85	14	14.8	16.7	38.6
	SD	0.5	2.28	0.64	2.06	3.4	3.51	7.65	13.9
Foot of Slope	M	0.76	3.87	3.59	9.15	14.3	16.2	13.8	38.4
	SD	0.65	2.81	0.64	1.87	1.81	3.95	1.55	3.84
Discharge zone		0.35	2.3	4.43	11.9	17.6	15.6	13	34.8
Tips									
Slope	M	0.2	1.72	3.1	8.89 a	14.5 a	13.3	14.2 b	44.1
	SD	0.25	0.62	1.37	4.14	6.21	3.8	2.92	15.8
Foot of Slope	M	0.34	2.3	4.45	12.9 ab	20.1 ab	16.7	12.9 ab	30.4
	SD	0.23	0.45	1.02	3.27	5.03	5.4	1.79	14.5
Discharge zone	M	0.26	2.76	5.78	16.0 b	22.9 b	17.8	9.73 a	24.8
	SD	0.21	1.08	2.3	4.95	3.42	1.3	2.2	10.2

Heterogeneous Distribution Of Pollutants

Through PCA, we tried to gain insight into the structure of the soil cap used to seal the landfill. In Figure 7A, it may be seen that the first axis, or component, is closely and positively linked to heavy metal and organic compound pollution although Na and F also appeared in this group of variables, and negatively related to soil fertility due to the presence of K and P. The second component was more related to soil salinity, represented by electrical conductivity, chlorides, sulfates, nitrates and nitrites. When organic components and the trace elements Hg and As were excluded, results failed to vary significantly and the first component continued to be positively and closely linked to the presence of heavy metals and negatively linked to that of K (Figure 7B). The second component, more related to salinity or electrical conductivity, this time was linked more to chlorides than the other anions.

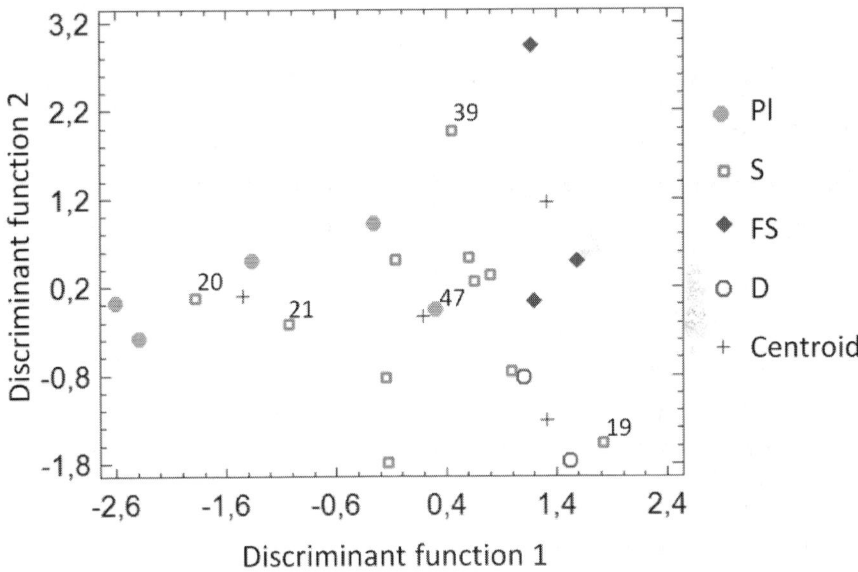

Figure 6. Representation of discriminant functions calculated with Mastersizer results of landfill soil samples, grouped in platforms (P), slopes (S), foots of slopes (FS) and discharge zones (D).

These findings confirm our previous results indicating that despite the uneven distribution of pollutants, at the most polluted sites all pollutants contribute to this contamination. The PCA plot of points on the new axes

(Figures 7C and 7D) serves to visually identify the sites showing highest heavy metal pollution as the landfill slopes and those with the greatest salinity as the rubble tips. The platforms emerged as the least polluted sites both in terms of heavy metals and salts contents.

The chemical analysis results reveal great heterogeneity in both the distributions and concentrations of pollutants. As an example of the complexity of the problem addressed, Zn concentrations range from 9 mg kg^{-1} to 23000 mg kg^{-1}; maximal Cd concentrations are 308 mg kg^{-1} (of which 85% represents the easily soluble fraction) and the maximal concentration of total hydrocarbons is 3408 mg kg^{-1}.

The spatial distributions of these factors determined using a Geographical Information System (ArcMap^{TM} software, v. 9.3.1., ESRI) are depicted in Figure 8.

UNDERSTANDING THE COMPLEX NATURE OF LANDFILL SOIL CAPS WITH THE VIEW OF RESTORING THE IMPACTS OF POLLUTION

The mountains of waste and rubble we have created are new landscape features that most often emerge in areas around cities. These scenarios can be viewed as laboratories for research into the environmental impacts of landfills that were capped without prior treatment of the deposited waste. Even considering that the restoration of degraded ecosystems is a systemic topic, the functionality of this epistemological approach arises from the fact that ecosystems are dynamic systems that evolve and co-evolve with human activity.

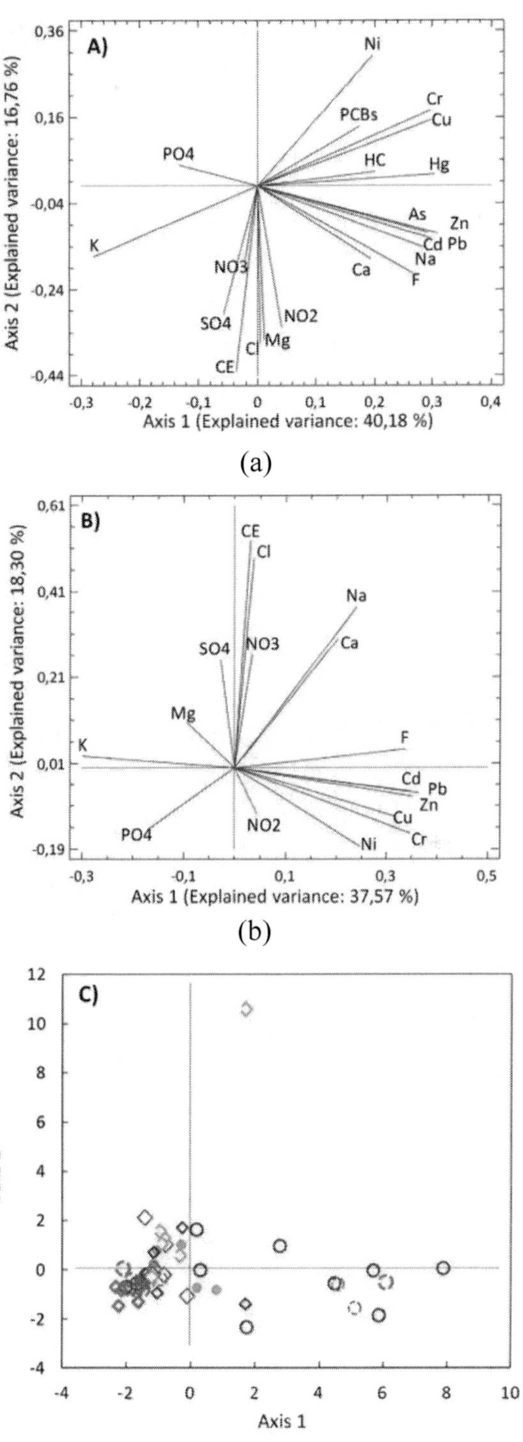

(a)

(b)

(c)

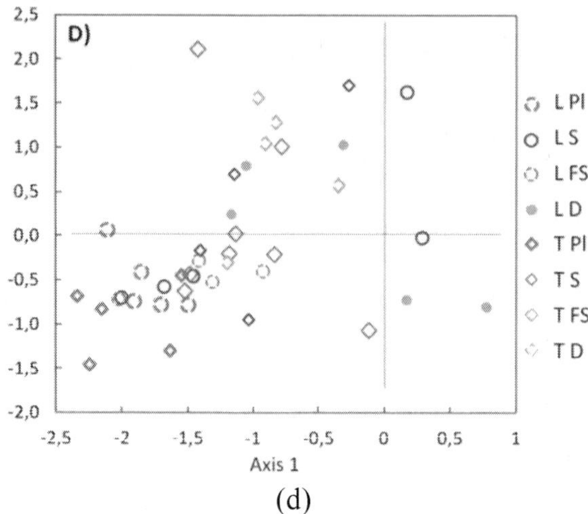

(d)

Figure 7. Principle components analysis of the soil chemical variables showing points appearing on the new coordinate axes. A) PCA of the whole set of variables, N = 29, B) PCA excluding As, Hg, hydrocarbons and PCBs increasing the number of cases to N = 52, C) representing the 52 points on the new axes created by the PCA in Figure B), D) expansion of plot C) from –2.5 to 1 abscissa and -2 to 2.5 ordinate. L, landfill proper; T, rubble tips; Pl, platforms; S, slopes; FS, foot of slopes; D, discharge zone.

The complexity of the problem faced arises from questions related to the secondary ecological succession (from the capping soil's seed bank), which interacts with the primary succession that is possible in this new ecosystem in the landscape. Besides restoring its impacts, efforts need to also focus on revegetating the landfill system itself.

Hence, these landfills may be considered a new type of ecosystem in which primary and secondary successions coincide. They are thus of great interest for ecological science since they provide a real scenario for investigating the measures we should install to restore a degraded and polluted ecosystem and help us identify the plant species related to their varied forms of pollution. This will enable researchers to select the most appropriate plant species for revegetation efforts rather than simply establishing a green cover once a landfill has been capped.

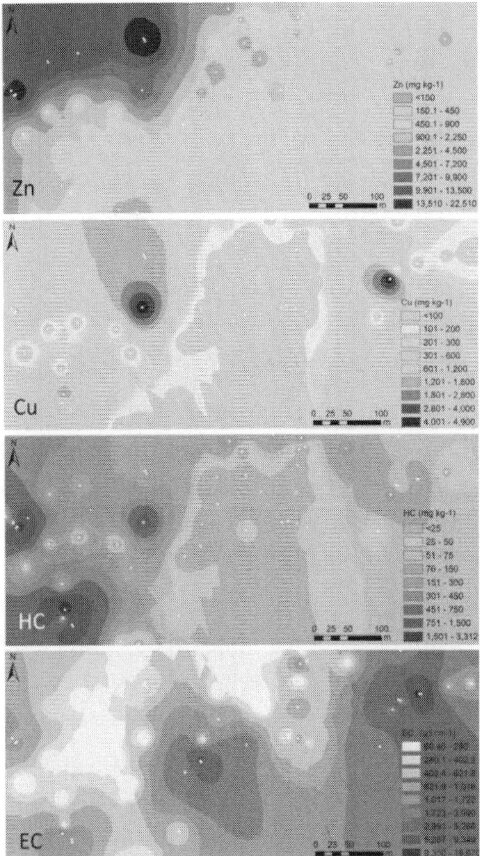

Figure 8. Spatial distribution of Zn, Cd, total hydrocarbons and electrical conductivity

The ecological theory that is most applicable to the restoration of the environmental impacts of capped landfills addresses the stress and ecological strategies of herbaceous species.

The classification of plant life cycle strategies described by Grime combines the stress intensity with the perturbation intensity [7, 28]. Thus, "competing species" are more appropriate for landfills with a low intensity of perturbation and stress, "ruderal" species adapt better to conditions of low stress and intense perturbation, and "stress-tolerants" are ideal for settings of intense stress and scarce perturbation. When both these factors are excessive, this approach is ineffective.

It should not be forgotten, however, that different types of ecosystem respond differently to a given perturbation, and vice-versa, that a given ecosystem can respond in many different ways to different perturbations. We also need to be aware of the vast environmental variability and randomness that exists along with other associated forms of uncertainty [29].

CONCLUSIONS

If a sealed landfill needs to be revegetated, it will be necessary to study the fertility of its soil cover, heavy metals and trace elements that can cause plant toxicity, salinity and organic compounds in the capping soil layer. The research methodology used in the landfill case study can be followed in other scenarios with a similar problem.

The analysis of all considered parameters and the heterogeneous distribution of pollutants indicate that a single-species cover should be avoided. It will be necessary to create a multispecies cover that will adapt to the heterogeneous distribution of the organic and inorganic pollutants present in capping soils and to the morphological features of the landfill's slopes.

From a scientific viewpoint, the scenario of the closed waste landfill has enabled the in depth study of what we have called the erosion-pollution binomial. This is the complex situation found in the capping soils of closed landfills in the Mediterranean setting. The plant species used for their revegetation should have the capacity to show an adequate response to this biome. To find such species, there is an urgent need for autecological studies and studies designed to assess native and commercial plant species that are able to adapt to these particular conditions. This is the reason why these results should not be extrapolated to other non-Mediterranean settings.

ACKNOWLEDGEMENTS

Authors acknowledge program P2009/AMB-1478 Community of Madrid Program (EIADES). MJGG was funded by the FPU fellowship (AP2008-02934) of Spain's Ministry of Education.

REFERENCES

1. Adarve MJ, Hernández AJ, Gil A, Pastor J. B, Zn, Fe and Mn content in four grassland species exposed to landfill leachates. Journal of Environmental Quality 1998;27: 1286-1293.
2. Adarve MJ. Análisis de la incidencia ambiental de vertederos de residuos sólidos urbanos en aguas, suelos y especies vegetales de zonas de descarga. PhD thesis. Universidad de Alcalá; 1993.
3. Cheng CY, Chu LM. Phytotoxicity data safeguard the performance of the recipient plants in leachate irrigation. Environmental Pollution 2007;145(1):195-202.
4. Fatta D, Papadopoulos A, Loizidou MA. Study of the landfill leachate and its impact on the groundwater quality of the greater area. Environmental Geochemistry and Health 1999;21: 175-190.
5. Gutiérrez-Ginés MJ. AJ Hernández. R Millán. J Pastor. Study of Hg in soil cover of sealed urban landfills and its toxic impact in Lupinus albus L. grown in landfill soils with high levels of this trace element. In: 9th Iberian and 6th Iberoamerican Congress on Environmental Contamination and Toxicology, CICTA2013, 1-4 July 2013, Valencia, Spain; 2013.
6. Hernández AJ, Adarve MJ, Gil A, Pastor J. Soil salination from landfill leachates: effects on the macronutrient content and plant growth of four grassland species. Chemosphere 1999;38: 1693-1711.
7. Hernández AJ, Adarve MJ, Pastor J. Some impacts of urban waste landfills on Mediterranean soils. Land Degradation & Development 1998;9: 21-33.
8. Hernández AJ, Bartolomé C, editors. Estudio multidisciplinar de vertederos sellados. Diagnóstico y pautas de recuperación. Alcalá de Henares: Servicio de Publicaciones de la Universidad de Alcalá; 2010.
9. Hernández AJ, Pastor J. Técnicas analíticas para el estudio de las interacciones suelo-planta. Henares, Revista de Geología 1989;3: 67-102.
10. Hernández AJ, Pastor J. Validated Approaches to Restoring the Health of Ecosystems Affected by Soil Pollution. In: Domínguez JB (ed.) Soil

Contamination Research Trends. New York: Nova Science Publishers; 2008. p51-72.

11. Hernández AJ, Pérez-Leblic MI, Bartolomé C, Rodríguez J, Álvarez J, Pastor J. Ecotoxicological diagnosis of a sealed municipal landfill. Journal of Environmental Management 2012;95: S50-S54.

12. Kalčíková G, Zagorc-Končan J, Zupančič M, Žgajnar Gotvajn A. Variation of landfill leachate phytotoxicity due to landfill ageing, Journal of Chemical Technology and Biotechnology 2012;87(9): 1349-1353.

13. Lakanen E, Ervio R. A comparison of eight extractants for the determination of plant available micronutrients in soils. Acta Agricultura Fennica 1971;123: 223-232.

14. Liphadzi MS, Kirkham MB. Physiological Effects of Heavy Metals on Plant Growth and Function. In: Huang B (ed.,) Plant-environment interactions. New York: Taylor & Francis; 2006. p243-269.

15. Millán R, Gamarra R, Schmid T, Sierra MJ, Quejido AJ, Sánchez DM, Cardona AI, Fernández M, Vera R. Mercury content in vegetation and soils of the Almadén mining area (Spain). Science of The Total Environment 2006;368(1): 79-87.

16. Mossop KF, Davidson CM. Comparison of original and modified BCR sequential extraction procedures for the fractionation of copper, iron, lead, manganese and zinc in soils and sediments. Analytica Chimica Acta 2003;478: 111–118.

17. Pastor J, Alía M, Hernández AJ, Adarve MJ, Urcelay A, Antón FA. Ecotoxicological studies on effects of landfill leachates on plants and animals in Central Spain. The Science of the Total Environment, 1993;140: 127-134.

18. Pastor J, Hernández AJ. Heavy metals, salts and organic residues in old solid urban waste landfills and surface waters in their discharge areas: Determinants for restoring their impact. Journal of Environmental Management 2012;95: S42-S49.

19. Pastor J, Hernández AJ. La restauración en sistemas con suelos degradados: estudios de casos en vertederos, escombreras y emplazamientos de minas abandonadas. In: Millán R, Lobo C (eds.) Contaminación de Suelos: Tecnologías para su recuperación. Madrid: CIEMAT; 2008. p539-560.

20. Pastor J, Urcelay A, Adarve MJ, Hernández AJ, Sánchez A. Aspects of contamination produced by domestic waste landfills on receiving waters in Madrid province. In: Nath B. et al. (eds.) Environmental Pollution. Science, Policy, Engineering. London: European Centre for Pollution Research; 1993. p254-261.

21. Pastor J, Urcelay A, Oliver S, Hernández AJ. Impact of Municipal Waste on Mediterranean Dry Environments. Geomicrobiology Journal 1993;11: 247-260.

22. RD 1310/1990. Real Decreto 1310/1990 de 29 de octubre, por el que se regula la utilización de los lodos de depuración en el sector agrario. Boletín Oficial del Estado del 1 de noviembre de 1990.

23. RD 9/2005. Real Decreto 9/2005 de 14 de enero, por el que se establece la relación de actividades potencialmente contaminantes del suelo y los criterios y estándares para la declaración de suelos contaminados. Boletín Oficial del Estado del 18 enero 2005.

24. Rivas-Martínez S, Diaz TE, Fernández-Gonzalez F, Izco J, Loidi J, Lousa M, Penas A. Vascular Plant Communities of Spain and Portugal. Addenda to the syntaxonomical checklist of 2001. Itinera Geobotanica 2002;15: 433-992.

25. Tatsi AA, Zouboulis AI. A field investigation of the quantity and quality of leachate from a municipal solid waste landfill in a Mediterranean climate (Thessaloniki, Greece). Advances Environmental Research 2002;6: 207-219.

26. Urcelai A. Estructura de sistemas herbáceos mediterráneos sometidos a la acción antrópica y posibles mecanismos de resiliencia. PhD thesis. Universidad de Alcalá; 1997.

27. Ursic KA, Kenkel NC, Larson DW. Revegetation dynamics of cliff faces in abandoned limestone quarries. Journal of Applied Ecology 1997;34: 289-303.

28. Walsh LM. Soil Society of America. Instrumental Methods for analysis of soils and plant tissue, vol VII. Soil Science Society of America, Wisconsin;1971.

29. Weber R, Watson A, Forter M, Oliaei F. Persistent organic pollutants and landfills. A review of past experiences and future challenges. Waste Management Research 2011;29: 107-121.

CITATION

Jesús Pastor, María Jesús Gutiérrez-Ginés, Carmen Bartolomé and Ana Jesús Hernández (2014). The Complex Nature of Pollution in the Capping Soils of Closed Landfills: Case Study in a Mediterranean Setting, InTech, DOI: 10.5772/57223.

CHAPTER 10

Optimization of Soil Erosion and Flood Control Systems in the Process of Land Consolidation

Miroslav Dumbrovsky[1] and Svatopluk Korsuň[1]

[1]Brno University of Technology, Faculty of Civil Engineering, Department of Landscape Water Management, Czech Republic

INTRODUCTION

Extreme hydrological phenomena of recent years have highlighted a well-known fact that it is necessary to pay greater attention to the problems of flood-prevention and soil erosion control on a large part of the Czech Republic. Case study areas are the most endangered territories. The case study area was selected as a case study mainly for its natural conditions and high risk of soil degradation and occurrence of flash floods. Relief, geomorphology, the present state of the complex system of soil properties, the types of agricultural farming practices and land use, are all contributing to accelerated soil erosion and runoff with all its negative impacts on the built-up areas.

The main soil degradation problems in the case study area are soil erosion caused by water, soil compaction and decline in organic matter. Soil erosion is fostered by i) soil degrading (intensive) farming practices such as up and down hill conventional tillage and other conventional agricultural operations on arable land, ii) frequent extreme hydrological events, and iii) a decreasing ability of soils for water retention (decline in organic matter and land conversion). Soil compaction is a problem due to intensive conventional farming on arable land (using heavy machinery).

The decline in organic matter results from the constant soil erosion process. Main causes of decline in organic matter are conventional farming practices without using manure and other organic matter. Decline in organic matter causes a decrease of natural crop productivity of soil and decreases yield.

Great runoffs occur on these areas and transform into flood waves in watercourses. Forest grounds are also affected, especially in case of unsuitable transport, wood cut and growth make-up. Solving of the problems of territory protection from unfavourable and damaging effects of overland water flow must therefore begin in catchments areas and particularly during any interference with landscape. Appropriate conservation measures are required to prevent and reduce runoff and soil degradation resulting from intensive agriculture. The adoption of the most appropriate practices and optimisation the farming conservation system it is necessary to carry out analyses and evaluations of the erosion rate and the basic characteristics of runoff in given sub-catchments. This system of evaluation provides information about erosion and runoff risks plots and serves for decision making regarding soil conservation and flood prevention measures. The success of the system of soil conservation depends on suitable technical assistance and support from responsible state organisations (Ministry of Agriculture and Ministry of Environment), sufficient sources of information as well as the ability and willingness of land users to adopt soil conservation measures. The main motivation for farmers to apply soil conservation measures is the economic motivation through financial subsidies along with penalties for farmers if they fail to comply with the rules of the funding program.

Nevertheless, when introducing a soil erosion control and flood prevention measures in a certain watershed, best management practices are mostly to be able decrease of erosion rate but unable to restrict a surface runoff substantially. For that reason it is necessary to apply a whole system of soil conservation measures. In places with long slopes technical and biotechnical soil erosion control practices (primarily of linear character) are necessary. These technical measures are broad base terraces and channels in case study area. These biotechnical measures together with the implementation of grassed courses of concentrated surface runoff (grassed waterways) create an appropriate network of new hydrolines in the watershed. Biotechnical line elements of soil erosion control serve as permanent barriers or obstacles for water runoff and are designed in order to determine, by their location, the ways of land management. Some technical and biotechnical measures could be suitable regarding their technical feasibility, economic efficiency and environmental effectiveness. The spatially and functionally limited soil-conservation system in a given teritorry offers spaces and lines in which it would be possible to locate territorial systems of ecological stability under certain conditions. Soil conservation and flood prevention practices, connected with territorial systems of ecological stability can be characterized as desirable anthropogenic landscape-forming elements. These would form the appearance of the landscape and significantly enhance natural processes in the region. They create suitable biological conditions in spite of the fact that they mostly do not meet qualitative and dimensional characteristics of biocentres and biocorridors.

Highly fragmented land ownership is prevalent in the area Biotechnical and technical soil conservation measures cannot be applied without respecting property rights. Integral parts of any project of soil erosion control (its basic network) are usually line elements for soil erosion control (broad base terraces and channels etc.), which run across individual owners' fields. Therefore it is necessary to identify every owner and discuss with the project and relevant proposals. The greatest interventions with agricultural landscape are land consolidation which, apart from other less important objectives, are designed to completely eliminate or at least partly limit unfavourable effects of runoff (especially soil erosion) and thus to become one of the most important elements of territory organisation and protection. Therefore it was found suitable to design the system of the soil and water conservation in the process of land consolidation in the Czech Republic.

Optimal spatial and functional delimitation of soil erosion control practices in the landscape is one of the basic steps in the plan of comprehenshive land consolidation, in addition to the implementation of a new network of field roads and landscape features enhancing ecological stability. Soil erosion control and flood prevention practices are included in the system of public facilities within the framework of the land consolidation process (where property relations are consistently solved).

The definition of land consolidation is from Act No. 139/2002 Coll., on Reparcelling and Land Authorities and amending Act No. 229/1991 Coll., on the Arrangement of Ownership titles to Land and other Agricultural Assets, as amended. The land consolidation processes in case study area have started in 2005.

This procedure gives solutions to the whole area, both from the aspect of a new land and ownership arrangement (Figure 2) and from the aspect of soil conservation and flood prevention and improvement of environment (Figure 1).

Figure 1. Soil conservation and Flood prevention system.

Figure 2. Parcel of owners before and after land consolidation.

Recently, the process of complex land consolidation in the Czech Republic has provided a unique opportunity for improving the quality of the environment and sustainability of crop production through better soil and water conservation. The current process of the land consolidation consists of the rearrangement of plots within a given territory, aimed at establishing the integrated land-use economic units, consistent with the needs of individual land owners and land users.

Integrated territory protection can be reached by controlling runoff by means of design of terraces as a soil erosion control measures. A number of mathematical models, mostly simulation ones, to solve water-management problems have been compiled, some of which include the option of exact mathematical optimization. A certain summary of these models, including their characteristics and application possibilities, were elaborated by Kos (1992). An interesting combination of the application of a simulation and optimization model technique in the elaboration of design of a particular water-management system was described by Major, Lenton et al. (1979), a three-model approach to solve water-management systems was used by Onta, Gupta and Harboe (1991). Benedini (1988) dealt more generally with the design and possible applications of these models. Most likely, an optimization model has not been designed, which would enable to attach territory protection and the measures to eliminate the amount and accumulation of runoff in catchments areas to solving water-management problems.

The created procedure is a universal tool which can be applied for any territory. It enables to find the most suitable combination of all possible alternatives of various erosion controls and flood protection measures under given conditions of each particular site. Such sites do not always have to be ground used for farming. They may also include in forest or urban areas or site arrays in various territories.

METHOD

The optimization process of designing the system of *integrated territory protection* (the *IOU* system) begins with the processing of the system of organisational, agrotechnical, biotechnical and technical measures at individual sites of the case study territory. It is necessary to derive hydrograms of direct runoff from extreme rainfall events for each of these variants. Then it is necessary to elaborate the variants of terraces and other conservation measures on all sections of watercourses and variant of designs of retention protection reservoirs. Not only rivers, streams and brooks are included into the watercourse category within this procedure but also sometimes passed watercourses such as terraces, grass infiltration belts or the lines of stabilisation of concentrated runoff waterways in valley lines.

A selection of the most suitable combination of all prepared variants is listed. With respect to the fact that it is necessary to find optimal dimensions for some of the system elements, there is usually a great number, in case of a continual solving even an infinite number, of possible combinations. It is therefore necessary to use an optimalized mathematical model to find the most suitable combination. This model was created on the basis of a mixed discrete programming (Korsuň et al., 2002, Dumbrovský et al., 2006). Its basic building stones are three generally formulated partial models: *A.* partial model of protective measures at individual sites of the case study area. *B.* partial model of a watercourse. *C.*partial model of a reservoir.

It is possible to shape an *optimization model of integrated territory protection* (*OMIOU*) from these partial models for any particular territory. The partial models are repeatedly inserted into the *OMIOU* as needed so as to exactly copy the modelled system structure. It is necessary to determine in advance one criterion or more simultaneously operating optimization criteria for each optimization function. A whole range of criteria can be determined for a given purpose. These can be taken from the sphere of economy but also from those of ecology, water-management, social etc. However it is necessary to define the most suitable criteria as far as quality is concerned but also to have a chance to quantify the values of each defined criterion. On top of that, it is necessary, in case of several simultaneously operating optimization criteria, to assign each criterion its adequate weight with which it will enter the solving process and which will support its effect on the result, so called a compromise solution in competition with the other criteria.

In creating the procedure of the *IOU* system proposal optimization in connection with the process of territory organisation a requirement of a maximal protection of inhabited and other areas with the exertion of minimal means was formulated for the solving process on the level of land consolidation as one of the suitable optimization criteria. It is a criterion consisting of three simultaneously operating partial economic, but at the same time water-management and socially aimed at their impacts. Criteria include:

- Minimization of the average annual damage (material damage: it is estimated that input requirements and conditions will not allow solutions which could lead to losses of human lives) originated by overland runoffs from rainfall events and then by their concentration in watercourses.

- Minimization of the average annual economic losses in farming production related to the realisation of proposed protective measures on arable land.
- Minimization of the average annual expenses (the sum of expenses for running and maintenance plus the amortization of the capital goods) of the proposed conservation measures.

Seeing that in most cases they are average annual values, quantified for example in thousands of CZK per year, these criteria can be assigned the same weights 1:1:1 in reflection.

The optimization mathematical model is a system of equations, which model a given system behaviour, the variables in the equation describe a system structure and the dimensions of its individual elements. Non-equations found in each model are transformed into equations by means of additional variables in the course of the model solving process, therefore the term *equation* is used only. The above mentioned partial models were created in the modelling and calculation system *GAMS* (*General Algebraic Modelling System*) in its general form (Charamza, 1993) so it can be used to model any integrated territory protection system. The nature of the solved problems implies that the defining process of all the variables used in the model as positive variables. They can be either continuous ones which are marked x herein after or binary ones (they can take on only 0 or 1 values) marked with the symbol x B. Other symbols are used to mark variables and coefficients. Activities proceeding in time must be modelled in the whole system according uniform timekeeping.

The partial model A is aimed at terraces and other biotechnical, agrotechnical, and organisation conservation measures in the catchments area of a certain watercourse. These measures are usually designed within land consolidation to decrease overland flow of rainfall events and thus to limit the effects of soil erosion and damage in inhabited territories. The various proposals of protective measures must be elaborated in each individual case before an optimization model is designed (pre-optimization) as pragmatically created systems of various, mutually complementary interventions with the individual catchments area elements. Such a partial catchments area element could be, for example, valley and slope area above one bank of a certain watercourse section in the range from the bank line to the interstream divide line.

The part of runoff from the design rainfall events which will not be caught by the system of catchments area protective measures (*residual runoff*) will concentrate in a particular watercourse and will create a design Q runoff or flood wave. The time T of passage of the design flood wave through a watercourse will be divided into r of equally long *time intervals* (*TI*); time t of the durance of one *TI* will thus be given by the relation $t = T / r$. For the individual *TI*s, partial volumes $w\,1$ of the design flood wave are then quantified, $i = 1, 2, ..., r$.

In case of the application of the above mentioned optimization criterion, the following indicators must be quantified for each pre-optimization processed variant of the protective measure set on a partial catchments area element:

- its estimated effect U expressed financially as an average annual level of damage on land, growth, buildings, roads etc. which will occur after the variant has been realised (*residual damage*),
- estimated average annual economical loss E in farming production related to the realisation of the proposed measures on arable land.
- realisation costs of a particular variant and its average annual own costs N,
- the amount of residual runoff Oi into a watercourse in the individual *TI*s.

These data represent input information for the partial model A. In the course of the optimization process, only one – optimal – variant with the most suitable indicators will be chosen from thus prepared variants of systems of protective measures for each partial catchments area element. Residual runoffs concentrating in a watercourse runway from the watercourse adjacent partial areas protected by optimal systems of measures will cause a gradual accretion of a flood wave passing through the watercourse. The protection from damage which could be caused by this flood wave will be provided by the protective measures on the watercourse and retention protective reservoirs as mentioned later (the partial models B and C).

Binary variables can be used for modelling of individual variants of protective measure systems in each of the partial catchments area elements in a discrete way. The total number of catchments area elements will be m. If, for example, n variants of protective measure systems of a $d^{\,th}$ catchments area element are modelled by relations to binary variables $x_{B1dp} \in \{0, 1\}$, $d = 1, 2,..., m$, $p = 1, 2,..., n$, the effects of these measure systems for this catchments area element can be write into the model using the following equations:

the equation of protective effects (residual damage)

$$x_{Ud} = \sum_{p} U_{dp} \cdot x_{B1dp}$$

$$(1)$$

the equation of economic damage

$$x_{Ed} = \sum_{p} E_{dp} \cdot x_{B1dp}$$

$$(2)$$

the equation of own costs

$$x_{Nd} = \sum_{p} N_{dp} \cdot x_{B1dp}$$

$$(3)$$

the equation of residual runoff, i.e. contribution of a $d^{\,th}$ catchments area element to the flood wave volume on a particular watercourse section in $i^{\,th}$ TI

$$x_{Oid} = \sum_{p} O_{idp} \cdot x_{B1dp}$$

$$(4)$$

Options

for $i = 1, 2,..., r$,

$d = 1, 2,..., m$,

$p = 1, 2, \ldots, n,$

where x_{Ud} is the total residual damage in a dth catchments area element,
$x_{E,Ud}$ is the total economic loss in a dth catchments area element,

x_{Oid} is the total residual runoff from a dth catchments area element in i^{th} TI.

Because only one of the protective measure system variants can enter the solving process, the following condition must be valid for the sum of all the binary variables of a dth catchments area element:

$$\sum_{p} x_{B1dp} = 1$$

(5)

The partial model B captures the passage of the design flood wave through the watercourse sections. The sections are either left in their present state, the optimization of a river bed or a contour furrow systems design (including the building of protective dams), or the reconstruction an earlier carried out adjustment or protective dams may be required. A watercourse section can also be a water or dry protective reservoir which will be modelled in a way mentioned in the partial model C description.

Flood damage that can occur is quantified for each watercourse section during its modelling. Further, runoffs from the section are calculated in the individual *TI*s of a flood wave passage. With respect to the overland flow from the initial section profile to the last one, it is necessary to determine a time shift which will affect collisions of flood waves on the main watercourse and at the mouths of its tributaries. The mean value of the runoff volume which can be found in a section (in a river bed or also in an inundation territory) in the course of ith *TI* is at the position of the basic section variable. The values of the other variables are related to this variable: the variables of the water flowing through the section, time of concentration, the level of flood damage in the section, and the level of runoff from the section. The courses of these non-linear functions are derived from the watercourse pre-optimization variant designs. They are replaced with linear function part by part in the optimization model. The formulation of particular equations is mentioned in Chapter 5.1 of Patera, Korsuň et al. (2002).

The partial model C is outlined for a designed multipurpose water reservoir with unknown capacities of spaces protective controllable x_{OO}, protective non-controllable $x\ ON$ and total $x\ V$. The necessary volumes of the spaces of dead storage $S \geq 0$ and active storage capacity $Z \geq 0$ are constant – these values result from other than protective requirements. The objective of analysis is to find its dimension which, respecting the requirement to create the spaces S and Z, with its protective spaces will ensure the reduction of culminated runoff from the reservoir to its optimal level during the passage of the design flood wave. In cases when the designed water reservoir has only a protective function, the value of the Z variable is zero; the values of both variables are zero $S = Z = 0$ for a dry protective reservoir.

The unknown volume of the total reservoir space is a variable, whose value which is limited from above by the maximal value $V\ max$ corresponding with the biggest realisable variant of the reservoir design during the pre-optimization solutions. From below it is limited by the minimal variant, still acceptable for practice, with the total volume $V\ min$.

We cannot forget a situation when building a reservoir will not be acceptable due to the used optimization criteria. It is therefore necessary to introduce a binary variable $x\ B2 \in \{0,\ 1\}$ into the set of variable values. If this variable has a zero value, the reservoir will not enter the solving process, if $x\ B2 = 1$, the entry of the reservoir into solving is cleared. Then the volume of the total reservoir space (without evaporation and percolation) must correspond with the following conditions

$$x_V = (S + Z) \cdot x_{B2} + x_{OO} + x_{ON}$$

(6)

$$V_{min} \cdot x_{B2} \leq x_V \leq V_m$$

(7)

The equations modelling the passage of the design flood wave through a dam profile, the calculations of the volumes of individual reservoir spaces and of necessary financial means are described in Chapter 5.1 of Patera, Korsuň et al. (2002). The partial model C can be also used for already an existing reservoir with a constant volume of the total space.

The model compilation from the fore mentioned partial elements in the presented form requires the introduction of a set of concrete coefficients and variables into the model for the model equation system to copy completely a particular system of *IOU*. These coefficients and variables should be derived from the pre-optimization processed background materials. In the case of non-standard requirements of an *IOU* system structure, it is necessary to introduce other equations to the model. Such new equations would capture these requirements. The model solving process in carried out on a computer by means of some of the *GAMS* system tools.

MATERIAL

To verify the function and potential of the already described optimization procedure, a system of integrated territory protection was chosen that was proposed within the framework of land consolidation on the case study area between the town of Hustopeče and the village of Starovice in the Czech Republic (see Figure 1). The declining ground in this region is mostly used as arable land. Overland flow is concentrated into its main waterway, which enters the residential parts of Hustopeče. Considerable, and frequently repeating damage, is caused by soil erosion on farm crops, sediment transport from arable land and especially by flooding parts of the town.

The proposed system of integrated protection of this farming territory and town is based on a system of technical-biotechnical, organisational and agrotechnical soil erosion control measures on arable land and of two conservation measures: 1. transfer of concentrated runoff from the drainage furrow or channel *K1* in the main valley line over the terrain into the adjacent valley line and creating a channel *K2* entering watercourse, 2. building a dry protective reservoir (polder) *P1* to catch parts of runoffs from the main valley line and another polder on the channel *K2* in the adjacent valley line above the village of Starovice.

The *IOU* system design for the given territory is based on the situation which would occur during a rainstorm with hundred-year periodicity (design rainfall). The protective measures with pre-optimization were designed in ten different variants, volume and cost (own costs) functions were derived for both the polders. It is estimated that *P1* polder filling, which is a side basin for the channel*K1*, will proceed through the channel side overfall. For the individual soil erosion control measure alternatives volumes and accumulation of overland runoffs, derived from the design rainfall, in the form of runoff hydrograms from two catchments areas: from the polder *P1* catchments and from the polder*P2* catchments. The passage of runoff waves through dam profiles of both polders takes from 510 minutes in the alternatives 1 and 2 to 195 minutes in the alternatives 9 and 10. It requires limiting the culmination water passages in the river beds below the two polders: below the *P1* this passage (runoff from the *P1*), which will enter the city sewerage system in Hustopeče, should not exceed 0.125 $m^3.s^{-1}$, below the *P2* the passage limit should be, with regard to the protection of Starovice, chosen at 1.5 $m^3.s^{-1}$ at most and in variants of 1.0 and 0.5 $m^3.s^{-1}$ to determine the effect of this passage size on the *IOU*system optimal solution.

The optimization model consists of 3,506 equations with the total of 1,673 structural variables, 539 of which are binary variables. The model objective function (optimization criterion) minimises the sum of average annual values of flood damages, economic losses and biotechnical measures and polders own expenses in the proportion of 1:1:1. It is ensured that only one protection system alternative can enter the *OMIOU* optimal solution in both the catchments areas, but it can be different for each of the catchments. These alternatives are marked as *A1* and a particular alternative number for the *P1* polder catchments, the *P2* polder was allocated symbol *2* in a similar way. The polders can enter the solution but they also do not have to. The runoff wave from the *P1* polder catchments may be partly or completely transferred into the *P2* polder. Permissible maximum of water depth in the *P1* polder is 5.0 m, it is 4.34 m in the *P2* polder. The *P1* polder low outlet dimensions (the inside diameter of outlet pipeline of a round shape) d = 200 mm, there is a possibility of choice from d = 200, 300, 400, 500 or 600 mm for the *P2* polder.

RESULTS AND DISCUSSION

The model function and behaviour were examined first in relation to the project research objectives. Then the possibilities of experimentation on the model of the designed system were tested (Korsuň et al., 2002). The optimization process was carried out with the three above listed values of admissible maximal runoff from the *P2* polder and then in an experimental way with various runoffs from both catchments areas: with real runoffs derived from hundred-year rain storms for the individual variants of conservation measures in both the catchments areas, and with fictive multiples of these runoffs.. Variants with other changes in input conditions (e.g. without the polders entering the solving process) were calculated for the same reason. The results of these solutions are not listed here. Optimal solutions of variants No. 1, 3 and 5 correspond with the real state of the input conditions. These solution results were derived from overland flows from the grounds and from three real values of admissible maximal runoff from the *P2* polder above Starovice. The results of following experiments on the optimization model have led to a number of interesting findings. However, the most important finding is the fact that the experimental locality can be protected as required without any interference into plant production conditions, i.e. without any (on site) economic loss on the produce only by conservation measures themselves: by draining overland and hypodermic runoffs through contour furrows and channels in the*P2* polder. This protective system design is valid only provided the applied optimization criterion is kept. The resulting design can be different in the case of any change to the criterion (e.g. the changes in the weights of the three used partial criteria) or in case of the application of a different criterion.

CONCLUSION

The results of the practical application of the optimization procedure in designing terraces and retention reservoirs within integrated territory protection verify its functionality and applicability. In cases when it is not clear in advance which of the potential torrential rainfall could be the most dangerous, the model will provide solutions with all chosen rainfalls types for the result to comply with the territory protection requirements.

The created model can be used to find either one optimal solution or, in case it is necessary to verify the position of the optimal solutions with the changes of some input conditions and requirements, more times in more versions with variables and coefficients modified by these changes. The possibility of multiple application of this model and to obtain a whole set of optimal solutions visualises much better the character and behaviour of the designed system in reactions to modifications of the input conditions and requirements and thus enables to improve significantly the process of making decisions about the design final shape.

A great advantage of the model lies in the general formulation of its components – partial models of conservation measures at individual sites of the experimental locality, watercourse and reservoir. This should enable its problem-free application for optimization design of integrated territory protection under any conditions and at any site.

REFERENCES

1. M. Benedini, 1988 Developments and possibilities of optimisation models Agric. Water Manag. 13 329 358
2. M. Dumbrovský, et al. 2003 Optimisation of the system of soil and water conservation for runoff minimizing in certain watershed in the process of land consolidation (in Czech) Final research report, NAZV-QC1292 VÚMOP, Praha
3. P. Charamza, et al. 1993 Modelling system GAMS (in Czech). MFF UK, Praha.
4. S. Korsuň, et al. 2002 Creating and verification of the model for optimisation of soil and water conservation (in Czech) Final research report, A01-NAZV-QC1292 FAST VUT, Brno
5. Z. Kos, 1992 Water management systems and their mathematical models during climate changes (in Czech) Vod. Hosp. 7 211 216
6. D. C. Major, R. L. Lenton, et al. 1979 Applied water resource systems planning. Prentice- Hall Inter., Inc., London.
7. P. R. Onta, A. D. Gupta, R. Harboe, 1991 Multistep planning model for conjunctive use of surface and groundwater resources Jour. Water Res. Plan. Manag. 6 662 678
8. A. Patera, J. Váška, J. Zezulák, V. Eliáš, S. Korsuň, et al. 2002 Floods: prognosis, water streams and landscape (in Czech) ČVUT / ČVVS Praha

CITATION

Miroslav Dumbrovsky and Svatopluk Korsun (2012). Optimization of Soil Erosion and Flood Control Systems in the Process of Land Consolidation, Research on Soil Erosion, Dr. Danilo Godone (Ed.), ISBN: 978-953-51-0839-9, InTech, DOI: 10.5772/50327.

Index

Linear dynamic analyses, 113, 115
linearity, 94